AF789310

Division Algebras:
Octonions,
Quaternions,
Complex Numbers
and the Algebraic Design of Physics

Mathematics and Its Applications

Managing Editor:

M. HAZEWINKEL

Centre for Mathematics and Computer Science, Amsterdam, The Netherlands

Volume 290

Division Algebras: Octonions, Quaternions, Complex Numbers and the Algebraic Design of Physics

by

Geoffrey M. Dixon
Brandeis University, U.S.A.

KLUWER ACADEMIC PUBLISHERS
DORDRECHT / BOSTON / LONDON

Library of Congress Cataloging-in-Publication Data

Dixon, Geoffrey M.
Division algebras : octonions, quaternions, complex numbers, and the algebraic design of physics / by Geoffrey M. Dixon.
p. cm. -- (Mathematics and its applications ; v. 290)
Includes index.

1. Algebra. 2. Mathematical physics. I. Title. II. Series: Mathematics and its applications (Kluwer Academic Publishers) ; v. 290.
QC20.7.A4D59 1994
512'.57--dc20 94-13948

ISBN 978-1-4419-4746-8

Published by Kluwer Academic Publishers,
P.O. Box 17, 3300 AA Dordrecht, The Netherlands.

Kluwer Academic Publishers incorporates
the publishing programmes of
D. Reidel, Martinus Nijhoff, Dr W. Junk and MTP Press.

Sold and distributed in the U.S.A. and Canada
by Kluwer Academic Publishers,
101 Philip Drive, Norwell, MA 02061, U.S.A.

In all other countries, sold and distributed
by Kluwer Academic Publishers Group,
P.O. Box 322, 3300 AH Dordrecht, The Netherlands.

Printed on acid-free paper

This printing is a digital duplication of the original edition.

All Rights Reserved
© 1994 Kluwer Academic Publishers. Second Printing 2002.
Softcover reprint of the hardcover 1st edition 1994
No part of the material protected by this copyright notice may be reproduced or utilized in any form or by any means, electronic or mechanical, including photocopying, recording or by any information storage and retrieval system, without written permission from the copyright owner.

Table of Contents

Preface

I don't know who Gigerenzer is, but he wrote something very clever that I saw quoted in a popular glossy magazine:

"Evolution has tuned the way we think to frequencies of co-occurances, as with the hunter who remembers the area where he has had the most success killing game."

This sanguine thought explains my obsession with the division algebras. Every effort I have ever made to connect them to physics - to the design of reality - has succeeded, with my expectations often surpassed. Doubtless this strong statement is colored by a selective memory, but the kind of game I sought, and still seek, seems to frowst about this particular watering hole in droves. I settled down there some years ago and have never felt like leaving.

This book is about the beasts I selected for attention (if you will, to render this metaphor politically correct, let's say I was a nature photographer), and the kind of tools I had to develop to get the kind of shots I wanted (the tools that I found there were for my taste overly abstract and theoretical). Half of this book is about these tools, and some applications thereof that should demonstrate their power. The rest is devoted to a demonstration of the intimate connection between the mathematics of the division algebras and the Standard Model of quarks and leptons with $U(1) \times SU(2) \times SU(3)$ gauge fields, and the connection of this model to 10-dimensional spacetime implied by the mathematics.

If you understand what I have written to this point, then there is a good chance you have the background to understand this book. As to agreeing with its philosophical premises, you probably won't. I've never yet found two mathematicians or physicists who agree on the connection between mathematics and physics, and its depth. However, such agreement is not required to appreciate the material presented.

I owe thanks to several people: to my wife, Suzanne Young of Harvard, whose work in archaeometry compelled me in some weird way to write the book; to my thesis advisor, Hugh Pendleton of Brandeis, for his advice, interest, and forbearance; and to Rafal Ablamowicz, Martin Cederwall, Pertti Lounesto, Corinne Manogue, Ian Porteous, and Tony Smith, all of whom at my instigation converged on Göteborg, Sweden, this last January (1994) to talk about division algebras, Clifford algebras, and physics. Each played a crucial role at this meeting, and the breadth and depth of their insights has greatly enriched this monograph.

Geoffrey Dixon, Waltham, Massachusetts, 21 February 1994.

1. Underpinnings.

This is a book of ideas. Most of them arise out of my fascination with the division algebras and their tantalizing potential for a deep connection to physics. Part of the book is a development of ideas that compellingly demonstrate that there is a connection. The rest is mathematics, applications of techniques originally developed for physics, but here applied to some of the special algebraic and geometric notions, like triality, associated with the division algebras. Inevitably many of these mathematical ideas will be shown to be intimately linked to the design of our physical reality.

1.1. The Argument.

Shortly after I began this monograph, in an attempt to crystallise the mathematical ideas that had motivated, and continue to motivate, this work, I wrote a series of essays, which I called Screeds, on Truth in physics, and how the mathematical ideas presented here conform to my notion of what that Truth should look like. The first two of these I call Gedanken Fantasies:

SCREED I

As you crack your eyes one morning your reason is assaulted by a strange sight. Over your bed, humming quietly, there floats a monitor, an ethereal, otherworldly screen on which is written a curious message: "I am the Screen of ultimate Truth. I am bulging with information and ask nothing better than to be allowed to impart it." In an attempt to establish the bonafides of the thing, or failing that, to drive it away that you may be about your business, you mumble a question.

"What is the proof of Fermat's last theorem?"

The screen flickers slightly, goes blank, and then there, in its righthand margin, under the heading "A Proof", is a proof.

So Fermat was wrong, you think. It does fit in the margin of a page. [Of course, subsequent mathematical history has rendered this piece of the screed outdated, but although now proven, or perhaps only nearly proven, there is still no proof that will fit in the margin of a page.]

Excitement builds as you realize that you, a theoretical physicist, have before you the last word in labor-saving devices. What should you ask it?

The screen flickers again, and a new message appears: "You have but one hour more." No doubt it has many calls on its time.

What should you ask?

Should you ask how the universe came into being? But what if this question has no answer in the context of ultimate Truth? Perhaps it were better to ask why, than how.

What about your own work on modelling physical reality? Do you suppose it could tell you if you are at least on the right path? Do you want to know?

Then a sobering thought occurs: there is no reason to believe you will understand the answers you are given. Do you believe sincerely that we have already at our disposal nearly all of the mathematical and physical concepts required to answer ultimate questions, and that, given time and some talented juggling of those concepts, a best fit to Truth may be found? What of the principles underlying the best fit to physical reality? What of the principles underlying the existence of physical reality? Do you believe in ultimate truths? Do you believe that evolution has given rise in the human cranium to an instrument capable of understanding?

Not wishing to disturb the person snoring softly at your side, you lean carefully forward and whisper a question.

What question?

The thought that the human brain, even taken collectively and over the entire past and future of the species, may ultimately prove inadequate to the task, occured to me as I wrote. Once I had argued the contrary view. But why should Nature produce a creature capable of stripping Nature bare? One needs first, perhaps, to believe that Nature has a plan, and that we fit prominently into it. And ultimately selfreferentiality may preclude any creature arising out of Nature from fully understanding the Nature whence it stems. But long before we encounter problems of that sort, we shall doubtless become bogged down by arbitrary and prejudiced human nature; it is our *lower* brain that drives us to strive, and which will finally keep us from succeeding.

The second Gedanken Fantasy follows from the first:

SCREED II

To airy nothing give a local habitation and a name.

Uruk lay back in bed and stared at the ceiling. The Monitor of ultimate Truth had departed, and he little doubted he would not see it again. He had asked it but one question: why does the universe exist? The monitor had responded with a faint crackling, what sounded like a weary sigh, and then the screen had filled with symbols and words for which Uruk had no context. Ignoring his pleas to cease, it used up his hour with notions too abstruse for Uruk to comprehend even in part. At the end of an hour it stopped and a message appeared:

"I am of course not done, but my time is up. I am sorry if you feel cheated by what you saw, for I know you can not have understood even a small fraction of the answer. But I have been strangely remiss if I have left you with the impression that the answer is not deep. Let me leave you with a thought. Those theorists accounted by you as great were humans of intelligence, but the magnitude of their accomplishments stems more from the paths they chose than it does from their mental powers. When all the world is mired even a single individual who finds a right path can accomplish seeming wonders. But right paths are harder to find than identify. If you find one you will experience a sudden lack of resistance, new realms of thought and application open rapidly and compellingly before you. To find one you must be guided by a right philosophy, by right principles, and luck. Without these, though all the greatest minds on earth should commit their energies to the search, no path from the mire will be found. Look about you and see if it is not so. Be prepared to unlearn when faced with discovery, to relearn in a new light."

Uruk watched the ceiling - it wasn't doing anything - and thought. Had he harbored any notion that the answer to his question might be religious, he was now disabused. In no conventional sense could the monitor's answer be so construed. He had been looking at ... what? Mathematics? Perhaps, but not as he knew it. Metamathematics? But of what else could the answer have consisted? Perhaps he had imagined he might be told the universe existed because of quantum bubbles, or something of that sort, but that would beg the question. Whence quantum? Whence bubble? Whence any of the form with which he was familiar? The answer had to lie in mathematics, but not as done by mathematicians. The universe of Uruk's experience is very specific. Conventional mathematics is full of generalizations, mostly of ideas drawn from the universe. The attention of physicists is focussed on the extant, with a hand in mathematics. Mathematicians keep a hand in physics, but gloved, lest their fingernails get dirty. One studies the tail, the other the trunk, of an elephant. Neither is even much aware of the vast body twixt the two ends that unifies them into a single science. In part, thought Uruk, it is the science of mathematical specialness. Is it an accident that of all the general notions of our mathematics, those we find that help us best to describe our universe are so often in some sense select? Could this be a guiding principle? Uruk shut his eyes and listened to the softly sussurating snoring of the sleeper sleeping soundly at his side. If custom reconciles us to everything, he thought, is custom binding us to the mire? Look about you said the monitor, and see if it is not so.

In another unfinished and more acerbic Screed, the Screen of ultimate Truth had more to say:

"The form of reality, that you as theorists are slowly beginning to understand, was before you were. The Truth that underlies it was before you were, and was before the invention of your mathematical formalisms. When you are gone, and a time will come when all of you are gone, the Truth will continue to exist, without the benefit of your formalisms, your definitions, theorems, your rigor.

"Do you believe that were I to explain as much of what I know as you could comprehend that you would recognize it, that you would say, oh yes, this is but an extension of what we have already done, and though the mathematics is broader, the principles deeper, I am not surprised? Do you think you have asked even a fraction of the questions you need to ask?"

The Screeds were intended to be read by others (which they were), the thought being that by posing such questions as those above, the readers' preconceptions might be softened, and their minds might be opened to what I subsequently had to say about the division algebras and physics. But we are most easily persuaded by ideas we are on the verge of discovering ourselves, and I have discovered that among physicists there is nothing remotely like a philosophical consensus. Still, a summation of my beliefs, or faith, if you will, just might prove a useful preamble.

27 million years ago there was a reality governed by the same mathematical/physical Truth that governs ours. The Truth was and is. It does not change. What we call mathematics is founded on a formal system. The system is a tool adapted to the human brain that it might see into this Truth a little, to give Truth a comprehensible shape. The tool, the system, is an invention. Gödel's famous undecidability theorem, for example, says nothing about Truth. It pertains to the system, and to its shortcomings. An eye may not see all of itself.

Truth is not an invention. With our mathematics we may discover parts of it, but in being undiscovered, even undiscoverable, Truth functions. 27 million years ago there was matter composed of quarks, and the quarks were subject to an $SU(3)$ symmetry. In a very real sense $SU(3)$ existed and functioned during all that vast history of the universe prior to our formulation of the notion of Lie group.

More significantly, the symmetry of the quarks was not, and is not, $SU(7)$, $SO(39)$, $Sp(12)$, or G_2. It was and is $SU(3)$. There are layers of Truth underneath this fact, which, in my opinion, force it to be so. They force observable space-time to have a (1,3)-pseudometric, and force isospin $SU(2)$ to break. They force reality to take the fundamental form it has taken. In showing that the alteration of a given feature, or collection of features, of our physical reality would lead to an inconsistent theory, we have not necessarily uncovered another layer of Truth. We should expect reality to be consistent, and any alteration of reality that leads to inconsistency to be impossible. Consistency, after all, is a concept that connects our mathematical system to reality. It is a check, both of the system, and of our application of it. I personally would be surprised, confronted with the Screen of ultimate Truth, if in telling me what it knows, it needed our mathematical axioms and physical principles, and that consistency was other than a consequence of something far more profound.

If one is to believe the literature most physicists think of mathematics as a collection of methods. Mathematics is accounted secondary to primary reality. It is laid on. Differential geometry applied to gravitation, algebraic topology applied to string theory, group representation theory applied to GUTs, are all mathematical methods.

Again judging from the literature, mathematicians are no closer to thinking that deep beneath their science, and physical science, there is a Truth to which both are profoundly tied. Mathematicians spend much time either in a selfreferential study of the system, or in using the system to generalize notions spawned by reality. So, for example, geometers do not study merely 3-space or 4-space, but n-space. Curiously they discovered that the most remarkable of all such geometries has dimension = 4. This fascinating result leads one to wonder if this is related to the fact that our reality's geometry has 4 apparent dimensions.

Conjecture of this sort - stemming from the observation of mathematical selectness in many aspects of the design of reality as we perceive it - arises out of the attractive notion that this design may not be arbitrary. It is an admission of faith that there is an undiscovered prereality Truth governing the features of this design.

While we may not be, perhaps never will or can be, in a position to formulate comprehensibly such a Truth, never-the-less faith in its existence can have a practical influence on research. The association of truth to beauty is such an influence, but this association is evidently one-way. Not all that is considered beautiful is true. Attesting to this is the plethora of often incompatible theoretical models each of whose adherents considers their baby beautiful.

Adding to beauty the notion of selectness results in a more usefully stringent guiding principle. It is this combination of qualities that so excites those theorists interested in applying the division algebras to physics. These algebras are certainly select. There are only four real normed division algebras: the reals, **R** (dim=1); the complexes, **C** (dim=2); the quaternions, **Q** (dim=4) (often denoted **H**, for Hamilton, their discoverer); and the octonions, **O** (dim=8) (sometimes called the Cayley algebra). And there the sequence ends. There are at least two ways to extend the sequence (the so-called Cayley-Dickson prescription is one, and I outline a more natural prescription in chapter 2), but none can result in more division algebras. There are only four.

As it turns out, in being division algebras they are also extremely generative. The Lie algebras $so(n;\mathbf{K})$ of nxn traceless, antihermitian matrices over the division algebra **K**, are just the sequences $so(n)$ ($\mathbf{K} = \mathbf{R}$), $su(n)$ ($\mathbf{K} = \mathbf{C}$), and $sp(n)$ ($\mathbf{K} = \mathbf{Q}$). (**O** is nonassociative, so it can not generate a Lie algebra in this way.) The Lie algebras of the five exceptional groups, G_2, F_4, E_k, k=6,7,8, are all associated with the octonions (see chapters 2, 7 and 8). The three parallelizable spheres, S^i, i=1,3,7, arise from **C**, **Q**, and **O**, and the sphere fibrations, $S^k \rightarrow S^{2k+1} \rightarrow S^{k+1}$, for k=0,1,3,7, are intimately linked to **R**, **C**, **Q**, and **O** (see chapter 7). The list goes on.

In physics most attention has been paid to their link to quantum mechanics, this interest principally spawned by the work of von Neumann and Jordan. Quantum mechanics must rest on a division algebra (**C** is the usual choice). In addition, the algebraic underpinnings of modern particle/field theory can be shown to arise from the division algebras. The Pauli algebra is isomorphic to $\mathbf{C} \otimes \mathbf{Q}$ (which is just the complexification of **Q**, or the quaternionization of **C**). Via exponentiation the imaginary element of **C** generates $U(1)$; those of **Q** generate $SU(2)$; and the automorphism group of **O**, G_2, has an $SU(3)$ subgroup that is the stability group of a fixed imaginary direction. $U(1) \times SU(2) \times SU(3)$ is, of course, the famous standard symmetry of quark and lepton theory.

This brings us to the final Screed, which will serve as a segue to what follows:

SCREED III

Lest too frequent repetition should blunt the force of the rant, herewith the deed.

The real normed division algebras are the reals, complexes, quaternions, and octonions (**R**, **C**, **Q**, **O**). There are no others. For this reason there are no other classes of classical Lie groups than orthogonal, unitary, symplectic, and exceptional, for they arise out of the division algebras. The Clifford algebras of the real pseudo-orthogonal geometries are all isomorphic to matrix algebras over **R**, **C**, and **Q** (if we use the left-adjoint algebras, $\mathbf{R}_L = \mathbf{R}$, $\mathbf{C}_L = \mathbf{C}$, $\mathbf{Q}_L = \mathbf{Q}$, $\mathbf{O}_L = \mathbf{R}(8)$, then the octonions can be included as well). Many (most?) of the special structures of analysis and topology are associated with the division algebras. The parallelizable spheres, for example, S^1, S^3, S^7, arise directly from exponentiation of the hypercomplex elements of **C**, **Q**, and **O**, and they are parallelizable because **C**, **Q**, and **O** are division algebras.

In mathematics the division algebras are fundamental and generative. Accepting that there is a subtle conjoining of mathematics and physics at depths beyond our present understanding, it would be impossibly strange did not the division algebras surface on the physics end of the pool equally fundamental and generative. Given that $\mathbf{R}\otimes\mathbf{C}\otimes\mathbf{Q} = \mathbf{R}_L\otimes\mathbf{C}_L\otimes\mathbf{Q}_L$ (tensor product) [Note: **R** is unnecessary in these tensor products over the reals. It was added only for emphasis.] is just the Pauli algebra in another guise; that $\mathbf{R}_L \otimes \mathbf{C}_L \otimes \mathbf{Q}_L \otimes \mathbf{O}_L$ (isomorphic to $\mathbf{C}(16)$) is to the increasingly important

(1,9)-geometry what the Pauli algebra is to the (1,3)-geometry, and that the individual groups of the standard symmetry, $U(1)$, $SU(2)$, $SU(3)$, have direct links to $\mathbf{C}$, $\mathbf{Q}$, and $\mathbf{O}$, it would require an effort of will to deny it is so.

But an even stronger case has been made. Let $\mathbf{T} = \mathbf{R} \otimes \mathbf{C} \otimes \mathbf{Q} \otimes \mathbf{O}$, $\mathbf{T}_L = \mathbf{R}_L \otimes \mathbf{C}_L \otimes \mathbf{Q}_L \otimes \mathbf{O}_L$. $\mathbf{T}$ is the object or spinor space of $\mathbf{T}_L$. $\mathbf{T}$ is 64-dimensional, so the $\mathbf{T}$ inner product has an $SO(64)$ symmetry. This is not gaugable, but the mechanism of the inner product has an $U(1) \times SU(2) \times SU(3)$ symmetry that is. With respect to this symmetry the elements of $\mathbf{T}$ transform precisely, down to the quantum numbers, as the direct sum of a family and antifamily of quark and lepton Pauli spinors. Moreover, this can happen in two ways, one precisely like left-chiral matter, right-chiral antimatter, and the other like right-chiral matter, left-chiral antimatter. So the mathematics accounts for the observed violation of parity. Moreover, spontaneous symmetry breaking can break $U(1) \times SU(2)$, while $SU(3)$ is necessarily exact. This is a mathematical result. Moreover, a phase arises from the mathematics that weak mixes the quarks, not the leptons, mathematics again in agreement with observation.

The spinors naturally link to (1,9)-space-time, in that $\mathbf{T}_L$ is the "Pauli" algebra of this geometry. With respect to this geometry leptons and quarks, even matter and antimatter, are indistinguishable. With respect to the derived $U(1) \times SU(2) \times SU(3)$ symmetry the spinors of $\mathbf{T}_L$ (ie., $\mathbf{T}$ itself) split into matter and antimatter, and into the individual quark-lepton, antiquark-antilepton parts. The (anti)matter projector of the algebra reduces (1,9) to (1,3) (one for matter, one for antimatter).

All of the algebraic features of the standard model are derived from $\mathbf{T}$ and its adjoints (internal $SU(2)$ is actually associated with $\mathbf{Q}_R$, the right adjoint algebra of $\mathbf{Q}$; the right adjoint algebra of $\mathbf{O}$ is the same as the left), including quarks, leptons, the standard symmetry, parity nonconservation, symmetry breaking, quantum numbers, family structure, family replication, and weak mixing. These are associated via the identification of the component spinors as of Pauli type with the appropriate (1,3)-space-time. And going beyond the experimentally confirmed, $\mathbf{T}$ and $\mathbf{T}_L$ give rise to a link to a (1,9)-space-time, and that link is precisely determined.

To dismiss the exact correspondence of $\mathbf{T}$-maths to observed reality as accidental or insignificant is wrong. Theories of which the standard model is a desired part but that are not based on $\mathbf{T}$-maths are wrong. Extended theories, like string theories, that require a (1,9)space-time, but link it to the standard model in a way different from that derived in $\mathbf{T}$-maths, are wrong.

They are wrong in the same way theories of the electron not based upon the Pauli and Dirac algebras and their spinors are wrong.

Uruk lay back with a sigh.

End of Screeds.

The words at the end of this Screed were written with generous strength, but I wished to avoid the mistake of the fellow who was so afraid of making a bad impression, he made no impression at all. In particular I wish now to avoid leaving the reader with the idea that I consider the division algebras to be a mathematical method, interchangeable with other methods, like the theories of matrix Clifford algebras and associated spinors, and Lie groups and associated multiplets and representations, by means of which the algebraic features of the standard model may be easily represented. Such a representation is sewn together by arbitrariness. There are infinitely many Clifford algebras and Lie groups, and having picked from these, there are infinitely many multiplets and representations. And when one has chosen a phenomenologically suitable collection, one has shot one's bolt. Laying on the methods of quantum field theory can give rise, of course, to many testable results, but in a broader sense the theory is a closed context with few or no pointers to deeper Truth. Even worse, like many scientific paradigms, this one is an ideological cage. It is filled with an infinite number of blocks with which to play, and amidst these blocks have been discovered some golden nuggets, but the rush that inevitably resulted from this discovery stranded a generation amidst boulders of basalt.

The division algebras not only represent the standard model, they come very close to deriving it. All the multiplets and algebra representations, even the allowable particle transitions, are fixed by the mathematics. What I shall present in this monograph is a development of this mathematics. The algebraic features of the standard model drop out en route. This is not an accident; this is a layer of Truth, and a doorway to deeper layers.

Finally, the physical consequences of the mathematics do not stop with the standard model. As I shall demonstrate, this theory beautifully connects the standard model to (1,9)-space-time. But waiting in the wings are all the little explored connections to those special topological and analytical structures associated with the division algebras. This is a seed waiting to germinate; a vast cavern waiting to be explored. Indeed, I do not wish to give the impression that I believe the division algebras to be a universal cure for all that ails us. Our symbolic mathematics is a tool, the microscope with which we attempt to see what we can not directly touch, and from which we are in all likelihood precluded from touching. But the Truth of Reality gives every indication of being accreted about the most select and beautiful of mathematical ideas, and the division algebras are just a part of the web of these.

This monograph begins in chapter 2 with a development of general division algebra theory with the focus on the octonions. The cyclic multiplication tables I employ are not new, but they are also not common in physical applications. They make calculations considerably easier than the conventional multiplication tables, which rely on the Cayley-Dickson process, which I will not cover. In section 2.6 I present a mathematical motivation for the approach which I hope will be viewed as compelling.

Heavily stressed throughout this monograph is the notion that with respect to the left, right, or two-sided adjoint actions of a division algebra on itself, the adjoint algebras behave like Clifford algebras, and the division algebras themselves, which are the objects of the adjoint actions, behave like spinor spaces. This idea is not new, but I am unaware of any treatment of the matter that makes the relation as obvious as that developed here.

In chapter 3 tensor products of the division algebras and their adjoints are considered, and inner products on the tensored division algebras (the spinor spaces) are developed. In the case of the tensor product of **C**, **Q**, and **O**, there is associated with this inner product a symmetry from which the standard symmetry of elementary particle theory is drawn.

In chapters 4 and 5 the mathematics developed in chapters 2 and 3 is applied to physics in the obvious way, as outlined already in this section. This includes identifying the quark and lepton fields, developing gauge fields from the internal symmetry, weak mixing, and the identification of the Higgs fields, whose action on the fermion fields is heavily restricted by the mathematics. These are used to spontaneously break the symmetry.

In chapter 6 it is shown that this entire theory can be seen to result from a theory of fermions based on a 10-dimensional space. With respect to this 10-dimensional space quarks and leptons, even matter and antimatter, are subsumed into one big hyperfield in which all are indistinguishable. Only by restricting the dependence on the extra six dimensions of the various identifiable quark and lepton bits of the hyperfield (these extra dimensions carry $SU(3)$ charges) does the theory dimensionally reduce to a more familiar form. The further consequences of this 10-dimensional theory are in the developmental stage.

At the end of chapter 6 I take a brief look at the link of the division algebras to string theory, especially focussing on applications of the octonions to string theory on 10-dimensional spacetime.

In chapter 7 some of the mathematical ideas used in the construction of the physical application of chapters 3 through 6 are applied to other problems, in particular to the parallelizable spheres, the Moufang identities, and $so(8)$ triality.

Chapter 8 continues the further application of the mathematics, beginning with an unconventional development of the Lie algebras of the Freudenthal-Tits Magic Square. The commutation relations of one of these algebras, the exceptional F_4, are explicitly developed as an example.

Several appendices are included, the use of which can make algebraic calculations easier.

In the remainder of this chapter some background material is developed, beginning with a review of Clifford algebra theory.

1.2. Clifford Algebras.

To any $m \leq n$ linearly independent vectors in an n-dimensional Euclidean space there corresponds an antisymmetric m-tensor (m-form) representing the oriented m-volume of the m-parallelopiped determined by the vectors. This m-form can be constructed via an antisymmetric "outer" product of the vectors.

Of equal importance is the symmetric "inner" product of vectors from which a scalar norm can be constructed.

Clifford algebras are associative algebras with identity that embody these outer and inner products in an elegant algebraic formalism [1]. Via the Dirac algebra, the foundation of the Dirac equation, and also the complexified Clifford algebra of (1,3)-spacetime (with pseudo-orthogonal metric signature $\{+---\}$), the importance of Clifford algebras to physics has been cemented and solidified.

The (universal) Clifford algebra of (p,q)-spacetime ($\mathbf{R}^{p,q}$), denoted $\mathbf{R}_{p,q}$, is generated by $n = p + q$ linearly independent elements, Υ_α, $\alpha = 1, ..., n$, satisfying

$$\boxed{\Upsilon_\alpha \Upsilon_\beta + \Upsilon_\beta \Upsilon_\alpha = 2\eta_{\alpha\beta}\epsilon,} \tag{1.1}$$

where

$$\eta_{\alpha\beta} = 0,\ \alpha \neq \beta,$$
$$\eta_{\alpha\alpha} = 1,\ 1 \leq \alpha \leq p,$$
$$\eta_{\alpha\alpha} = -1,\ p+1 \leq \alpha \leq p+q$$

(ϵ = identity). Therefore, if $x = \sum_\alpha x^\alpha \Upsilon_\alpha$ and $y = \sum_\alpha y^\alpha \Upsilon_\alpha$ are algebraic representations of $\mathbf{R}^{p,q}$ vectors in $\mathbf{R}_{p,q}$, then

$$\boxed{\tfrac{1}{2}(xy + yx) = \sum_{\alpha\beta} x^\alpha \eta_{\alpha\beta} y^\beta \epsilon,} \tag{1.2}$$

where $\sum_{\alpha\beta} x^\alpha \eta_{\alpha\beta} y^\beta$ is the $\mathbf{R}^{p,q}$ inner product of the vectors with components $\{x^\alpha\}$ and $\{y^\alpha\}$.

The elements Υ_α are a basis for the 1-vector subset of $\mathbf{R}_{p,q}$ (when p=1 the values of the index α will be taken from $\{0, 1, ..., n-1\}$). The 2-vector subset is spanned by the $\begin{pmatrix} n \\ 2 \end{pmatrix} = \frac{n!}{2!(n-2)!}$ distinct

$$\boxed{\Upsilon_{\alpha\beta} = \tfrac{1}{2}(\Upsilon_\alpha\Upsilon_\beta - \Upsilon_\beta\Upsilon_\alpha).} \tag{1.3}$$

If x and y are defined as above, then

$$\begin{array}{rcl} \tfrac{1}{2}(xy - yx) & = & \tfrac{1}{2}\sum_{\alpha\beta}(x^\alpha y^\beta \Upsilon_\alpha \Upsilon_\beta - y^\beta x^\alpha \Upsilon_\beta \Upsilon_\alpha) \\ & = & \sum_{\alpha\beta} x^\alpha y^\beta \Upsilon_{\alpha\beta} \\ & = & \sum_{\alpha\beta} \tfrac{1}{2}(x^\alpha y^\beta - y^\alpha x^\beta)\Upsilon_{\alpha\beta} \end{array} \tag{1.4}$$

where $\frac{1}{2}(x^\alpha y^\beta - y^\alpha x^\beta)$ in the $(p = n,\ q = 0)$ or $(p = 0,\ q = n)$ case are the components of the oriented 2-form representing the area (2-volume) of the parallelogram determined by the two vectors.

Likewise, a basis for the space of general m-vectors consists of the $\begin{pmatrix} n \\ m \end{pmatrix}$ distinct elements of the form

$$\Upsilon_{\alpha_1\alpha_2...\alpha_m} = \Upsilon_{\alpha_1}\Upsilon_{\alpha_2}...\Upsilon_{\alpha_m}, \tag{1.5}$$

where $\alpha_1, \alpha_2, ..., \alpha_m$, are distinct (in which case this element is antisymmetric in the indices).

A "universal" Clifford algebra satisfies

$$\Upsilon_1\Upsilon_2...\Upsilon_n = \Upsilon_{12...n} \neq \epsilon. \tag{1.6}$$

This element is called the pseudoscalar of the Clifford algebra. Unless otherwise specified the term Clifford algebra will mean universal Clifford algebra. Only certain (p,q)-spacetimes admit nonuniversal Clifford algebras. Some of these will appear in chapter 8. Universal Clifford algebras are 2^n-dimensional, while the nonuniversal are 2^{n-1}-dimensional.

The subspace of 2-vectors of $\mathbf{R}_{p,q}$ is easily shown to close under commutation, and in this case it is isomorphic to the Lie algebra $so(p,q)$. In particular, the set of all

$$U = exp(\theta^{\alpha\beta}\Upsilon_{\alpha\beta})$$

form a Lie group under multiplication. If $x = x^\alpha \Upsilon_\alpha$, then the action

$$x \rightarrow UxU^{-1} \tag{1.7}$$

results in an $SO(p,q)$ rotation on the components $\{x^\alpha\}$ of x, and this rotation leaves the bilinear form 1.2 invariant.

This same element U, if acting on the spinor space of $\mathbf{R}_{p,q}$ (a column vector space upon which a matrix representation of $\mathbf{R}_{p,q}$ naturally acts; the concept is further developed on the following pages) is an element of $Spin(p,q)$, which defines what this means. $Spin(p,q)$ is a double covering of $SO(p,q)$.

There is a well-known table of isomorphisms between the Clifford algebras $\mathbf{R}_{p,q}$ and matrix algebras over $\mathbf{R}$, $\mathbf{C}$, or $\mathbf{Q}$. I'll present the table in a slightly unconventional way, splitting it into two tables, one for $p+q=n$ even, and one for $p+q=n$ odd.

Clifford Algebras: n even

n=	0	2	4	6	8	10
p-q						
10						**R**(32)
8					**R**(16)	**R**(32)
6				**Q**(4)	**Q**(8)	**Q**(16)
4			**Q**(2)	**Q**(4)	**Q**(8)	**Q**(16)
2		**R**(2)	**R**(4)	**R**(8)	**R**(16)	**R**(32)
0	**R**	**R**(2)	**R**(4)	**R**(8)	**R**(16)	**R**(32)
-2		**Q**	**Q**(2)	**Q**(4)	**Q**(8)	**Q**(16)
-4			**Q**(2)	**Q**(4)	**Q**(8)	**Q**(16)
-6				**R**(8)	**R**(16)	**R**(32)
-8					**R**(16)	**R**(32)
-10						**Q**(16)

There are some obvious patterns. In particular note the

...RRQQRRQQ...

pattern of the columns, and along the constant p and constant q diagonals in the n even table.

Clifford Algebras: n odd

p-q \ n=	1	3	5	7	9	11
11						$\mathbf{C}(32)$
9					${}^2\mathbf{R}(16)$	${}^2\mathbf{R}(32)$
7				$\mathbf{C}(8)$	$\mathbf{C}(16)$	$\mathbf{C}(32)$
5			${}^2\mathbf{Q}(2)$	${}^2\mathbf{Q}(4)$	${}^2\mathbf{Q}(8)$	${}^2\mathbf{Q}(16)$
3		$\mathbf{C}(2)$	$\mathbf{C}(4)$	$\mathbf{C}(8)$	$\mathbf{C}(16)$	$\mathbf{C}(32)$
1	${}^2\mathbf{R}$	${}^2\mathbf{R}(2)$	${}^2\mathbf{R}(4)$	${}^2\mathbf{R}(8)$	${}^2\mathbf{R}(16)$	${}^2\mathbf{R}(32)$
-1	$\mathbf{C}$	$\mathbf{C}(2)$	$\mathbf{C}(4)$	$\mathbf{C}(8)$	$\mathbf{C}(16)$	$\mathbf{C}(32)$
-3		${}^2\mathbf{Q}$	${}^2\mathbf{Q}(2)$	${}^2\mathbf{Q}(4)$	${}^2\mathbf{Q}(8)$	${}^2\mathbf{Q}(16)$
-5			$\mathbf{C}(4)$	$\mathbf{C}(8)$	$\mathbf{C}(16)$	$\mathbf{C}(32)$
-7				${}^2\mathbf{R}(8)$	${}^2\mathbf{R}(16)$	${}^2\mathbf{R}(32)$
-9					$\mathbf{C}(16)$	$\mathbf{C}(32)$
-11						${}^2\mathbf{Q}(16)$

The pattern across the rows is clear in both tables. In the second table, for example, ${}^2\mathbf{R}(2)$ is the algebra of matrices of the form

$$\begin{bmatrix} X & 0 \\ 0 & Y \end{bmatrix}, \tag{1.8}$$

where $X, Y \in \mathbf{R}(2)$, and 0 is the 2×2 zero matrix.

CLIFFORD ALGEBRA $\mathbf{R}_{3,1}$

Of particular interest to physicists, and to the application to be developed in chapters 3 to 6, are the spacetimes $\mathbf{R}^{1,3}$ and $\mathbf{R}^{3,1}$, and $\mathbf{R}^{1,9}$ and $\mathbf{R}^{9,1}$. Consider first the Clifford algebra

$$\mathbf{R}_{3,1} \simeq \mathbf{R}(4). \tag{1.9}$$

We can represent the generators by the matrices

$$\Upsilon_1 = \begin{bmatrix} 1 & \cdot & \cdot & \cdot \\ \cdot & 1 & \cdot & \cdot \\ \cdot & \cdot & -1 & \cdot \\ \cdot & \cdot & \cdot & -1 \end{bmatrix}, \quad \Upsilon_2 = \begin{bmatrix} \cdot & \cdot & 1 & \cdot \\ \cdot & \cdot & \cdot & 1 \\ 1 & \cdot & \cdot & \cdot \\ \cdot & 1 & \cdot & \cdot \end{bmatrix},$$
$$\Upsilon_3 = \begin{bmatrix} \cdot & \cdot & \cdot & 1 \\ \cdot & \cdot & -1 & \cdot \\ \cdot & -1 & \cdot & \cdot \\ 1 & \cdot & \cdot & \cdot \end{bmatrix}, \quad \Upsilon_4 = \begin{bmatrix} \cdot & \cdot & \cdot & 1 \\ \cdot & \cdot & 1 & \cdot \\ \cdot & -1 & \cdot & \cdot \\ -1 & \cdot & \cdot & \cdot \end{bmatrix}. \tag{1.10}$$

In this representation

$$\Upsilon_{1234} = \begin{bmatrix} \cdot & \cdot & -1 & \cdot \\ \cdot & \cdot & \cdot & 1 \\ 1 & \cdot & \cdot & \cdot \\ \cdot & -1 & \cdot & \cdot \end{bmatrix}, \tag{1.11}$$

The elements of $\mathbf{R}_{3,1}$ can be viewed as endomorphisms of the real vector space of 4×1 matrices, which are in this context called the spinors of the Clifford algebra. Nature, however, seems unimpressed with this scheme, prefering instead the complexification of the Clifford algebra and its spinors. In particular, the Dirac algebra, upon which the physics of fermions is based, is the complexification of $\mathbf{R}_{3,1}$, and its spinors are 4×1 over $\mathbf{C}$ (Dirac spinors), instead of 4×1 over $\mathbf{R}$ (Majorana spinors). The addition of the complex unit i allows the construction of the following more interesting and physically more relevant representation of the generators of $\mathbf{R}^{3,1}$, which we now denote by the more conventional γ_α, $\alpha = 1,2,3,4$:

$$\gamma_1 = \begin{bmatrix} \cdot & \cdot & \cdot & 1 \\ \cdot & \cdot & 1 & \cdot \\ \cdot & 1 & \cdot & \cdot \\ 1 & \cdot & \cdot & \cdot \end{bmatrix}, \quad \gamma_2 = \begin{bmatrix} \cdot & \cdot & \cdot & -i \\ \cdot & \cdot & i & \cdot \\ \cdot & -i & \cdot & \cdot \\ i & \cdot & \cdot & \cdot \end{bmatrix},$$
$$\gamma_3 = \begin{bmatrix} \cdot & \cdot & 1 & \cdot \\ \cdot & \cdot & \cdot & -1 \\ 1 & \cdot & \cdot & \cdot \\ \cdot & -1 & \cdot & \cdot \end{bmatrix}, \quad \gamma_4 = \begin{bmatrix} \cdot & \cdot & 1 & \cdot \\ \cdot & \cdot & \cdot & 1 \\ -1 & \cdot & \cdot & \cdot \\ \cdot & -1 & \cdot & \cdot \end{bmatrix}, \tag{1.12}$$

where now all the γ_α are 2×2 block off-diagonal, and the $\gamma_{\alpha\beta}$, which generate the Lorentz group $SO(3,1)$, are block diagonal, as is the pseudoscalar

$$\gamma_{1234} = \begin{bmatrix} -i & \cdot & \cdot & \cdot \\ \cdot & -i & \cdot & \cdot \\ \cdot & \cdot & i & \cdot \\ \cdot & \cdot & \cdot & i \end{bmatrix}. \tag{1.13}$$

From this we define

$$\gamma_5 = i\gamma_{1234} = \begin{bmatrix} 1 & \cdot & \cdot & \cdot \\ \cdot & 1 & \cdot & \cdot \\ \cdot & \cdot & -1 & \cdot \\ \cdot & \cdot & \cdot & -1 \end{bmatrix}. \tag{1.14}$$

(γ_5 anticommutes with γ_α, $\alpha = 1,2,3,4$, and it is not an element of $\mathbf{R}_{3,1}$, the real Clifford algebra. Hence the set $\{\gamma_1, \gamma_2, \gamma_3, \gamma_4, \gamma_5\}$ can be viewed as a basis for the 1-vector subspace of $\mathbf{R}_{4,1} \simeq \mathbf{C}(4)$.) Nature employs γ_5 in the following way. Let $\varphi = \begin{bmatrix} \varphi_1 \\ \varphi_2 \\ \varphi_3 \\ \varphi_4 \end{bmatrix}$ be a Dirac spinor, and define

$$\varphi_\ell = \frac{1}{2}(1+\gamma_5)\varphi = \begin{bmatrix} \varphi_1 \\ \varphi_2 \\ 0 \\ 0 \end{bmatrix}, \text{ and } \varphi_r = \frac{1}{2}(1-\gamma_5)\varphi = \begin{bmatrix} 0 \\ 0 \\ \varphi_3 \\ \varphi_4 \end{bmatrix}. \tag{1.15}$$

These are the so-called left and right (chiral) Weyl spinors of which the Dirac spinor φ is composed, and the $\frac{1}{2}(1 \pm \gamma_5)$ are the chiral projection operators.

Note that because the generators of the Lorentz group are block diagonal, Lorentz rotations map the spaces of left/right Weyl spinors to themselves. Note also that without the complex unit i, γ_5 is not constructable.

Weyl spinors, instead of Dirac spinors, are the indecomposible bits upon which fermion physics is based. For example, the left and right chiral parts of the electron's Dirac spinor interact differently with the gauge fields arising from the electroweak gauge symmetry, $U(1) \times SU(2)$. In particular, $SU(2)$ gauge fields arise as though the lefthanded electron and its associated lefthanded neutrino are an $SU(2)$ doublet, while the righthanded electron interacts as though a singlet. Its associated righthanded neutrino is often assumed not to exist at all. So although mathematically the block off-diagonal representation of the γ-matrices in 1.12 is not singled out, a representation of this sort is singled out by the physics. Because the left and right chiral projections of spinor fields corresponding to real fermionic particles interact differently with some gauge bosons, nature requires the complexification of the $\mathbf{R}_{3,1}$ Clifford algebra, for without the complex unit i, the chiral projectors $\frac{1}{2}(1+\gamma_5)$ can not be constructed.

PAULI ALGEBRA

More interestingly, we see that the Dirac algebra is not so much 4×4 over $\mathbf{C}$ as it is 2×2 over $\mathbf{P}$=Pauli algebra; that $\mathbf{P}$ more properly constitutes the parts of which the Dirac algebra is composed.

The Pauli algebra

$$\mathbf{P} \simeq \mathbf{R}_{3,0}, \tag{1.16}$$

and its basis of generating 1-vectors is conventionally represented by

$$\sigma_1 = \begin{bmatrix} 0 & 1 \\ 1 & 0 \end{bmatrix}, \sigma_2 = \begin{bmatrix} 0 & -i \\ i & 0 \end{bmatrix}, \sigma_3 = \begin{bmatrix} 1 & 0 \\ 0 & -1 \end{bmatrix}. \tag{1.17}$$

Let

$$\sigma_0 = \begin{bmatrix} 1 & 0 \\ 0 & 1 \end{bmatrix}, \tag{1.18}$$

then

$$\gamma_j = \begin{bmatrix} 0 & \sigma_j \\ \sigma_j & 0 \end{bmatrix}, j = 1,2,3, \ \gamma_4 = \begin{bmatrix} 0 & \sigma_0 \\ -\sigma_0 & 0 \end{bmatrix}. \tag{1.19}$$

The linear subspace of $\mathbf{P}$ with basis $\{\sigma_0,\ i\sigma_j, j = 1,2,3\}$ is a subalgebra of $\mathbf{P}$ isomorphic to the quaternion algebra $\mathbf{Q}$

$$(\sigma_0 \to 1,\ i\sigma_j \to -q_j, j = 1,2,3.)$$

Hence,

$$\boxed{\mathbf{P} \simeq \mathbf{C} \otimes \mathbf{Q}.} \tag{1.20}$$

CLIFFORD ALGEBRA $\mathbf{R}_{1,3}$

The Clifford algebra $\mathbf{R}_{1,3} \simeq \mathbf{Q}(2)$, and we may represent the 1-vector basis by

$$\Upsilon_j = \begin{bmatrix} 0 & q_j \\ q_j & 0 \end{bmatrix}, j = 1, 2, 3, \ \Upsilon_4 = \begin{bmatrix} 1 & 0 \\ 0 & -1 \end{bmatrix}, \tag{1.21}$$

so that

$$\Upsilon_{1234} = \begin{bmatrix} 0 & 1 \\ -1 & 0 \end{bmatrix} \tag{1.22}$$

This algebra can be viewed as an algebra of endomorphisms on the space of 2×1 matrices over $\mathbf{Q}$, which is an unconventional spinor space for $\mathbf{R}_{1,3}$, being neither real nor complex. It is interesting in that the algebra of left actions of $\mathbf{Q}(2)$ on this spinor space is incomplete, and is only completed by including right actions of $\mathbf{Q}$ on the space of column spinors. This algebra of right actions commutes with the larger algebra of left actions, and is in that sense "internal" with respect to the algebra of left actions, which, in representing the Clifford algebra of the relevant Minkowski geometry, are considered "external". In that the set of unit quaternions is the Lie group $SU(2)$, which in the context just outlined could be viewed as an internal symmetry, it is curious that this representation of $\mathbf{R}_{1,3}$ has never been employed (although something similar will be employed in this monograph).

In complexifying $\mathbf{R}_{1,3}$ we are again able to define a chiral projection operator. In particular redefine in this new context,

$$\gamma_0 = \begin{bmatrix} 0 & 1 \\ 1 & 0 \end{bmatrix}, \ \gamma_j = \begin{bmatrix} 0 & -iq_j \\ iq_j & 0 \end{bmatrix}, j = 1, 2, 3, \tag{1.23}$$

where now we use the index 0 to denote the time direction. Given this basis,

$$\gamma_{0123} = \begin{bmatrix} -i & 0 \\ 0 & i \end{bmatrix}, \tag{1.24}$$

and from this we define

$$\gamma_5 = i\gamma_{0123} = \begin{bmatrix} 1 & 0 \\ 0 & -1 \end{bmatrix}. \tag{1.25}$$

As before, γ_5 together with the $\gamma_\mu, \mu = 0,1,2,3$, form a 1-vector basis for a larger Clifford algebra, in this case $\mathbf{R}_{2,3}$.

This is an unconventional representation still, involving as it does the quaternions. The more conventional representation is

$$\gamma_0 = \begin{bmatrix} 0 & \sigma_0 \\ \sigma_0 & 0 \end{bmatrix}, \; \gamma_j = \begin{bmatrix} 0 & \sigma_j \\ -\sigma_j & 0 \end{bmatrix}, j = 1,2,3. \tag{1.26}$$

This has the conventionally perceived advantage that its spinor space is the space of 4×1 complex column matrices, while the spinor space of the representation 1.23 is the space of 2×1 column matrices over $\mathbf{C} \otimes \mathbf{Q}$. That the representation 1.23 has an associated inherent internal symmetry $U(1) \times SU(2)$ is largely unrecognized, but a detailed study of this will follow in subsequent chapters.

CLIFFORD ALGEBRA $\mathbf{R}_{1,9}$

Both $\mathbf{R}_{1,9}$ and $\mathbf{R}_{9,1}$ are isomorphic to $\mathbf{R}(32)$, and they can be similarly analysed. We will develop a representation of $\mathbf{R}_{1,9}$.

Let

$$\epsilon = \begin{bmatrix} 1 & 0 \\ 0 & 1 \end{bmatrix}, \alpha = \begin{bmatrix} 1 & 0 \\ 0 & -1 \end{bmatrix},$$

$$\beta = \begin{bmatrix} 0 & 1 \\ 1 & 0 \end{bmatrix}, \omega = \begin{bmatrix} 0 & 1 \\ -1 & 0 \end{bmatrix} \tag{1.27}$$

be a basis for $\mathbf{R}(2)$. Define, for example, in $\mathbf{R}(4)$,

$$\beta \otimes \alpha = \begin{bmatrix} 0 & \alpha \\ \alpha & 0 \end{bmatrix}, \tag{1.28}$$

the tensor product of β and α, where the first matrix, β, determines the block structure, and the second, α, what goes in the blocks. Therefore,

$$\alpha \otimes \beta = \begin{bmatrix} \beta & 0 \\ 0 & -\beta \end{bmatrix}. \tag{1.29}$$

This easily generalizes to tensor products of arbitrary length, and in this manner we can construct any element of any $\mathbf{R}(2^k)$. In particular, define in $\mathbf{R}(32)$,

$$\begin{aligned}
\Upsilon_0 &= \beta.\epsilon.\epsilon.\epsilon.\epsilon \\
\Upsilon_1 &= \omega.\alpha.\epsilon.\epsilon.\epsilon \\
\Upsilon_2 &= \omega.\beta.\beta.\beta.\beta \\
\Upsilon_3 &= \omega.\beta.\alpha.\alpha.\alpha \\
\Upsilon_4 &= \omega.\beta.\epsilon.\alpha.\beta \\
\Upsilon_5 &= \omega.\beta.\beta.\epsilon.\alpha \\
\Upsilon_6 &= \omega.\beta.\alpha.\beta.\epsilon \\
\Upsilon_7 &= \omega.\omega.\omega.\alpha.\beta \\
\Upsilon_8 &= \omega.\omega.\beta.\omega.\alpha \\
\Upsilon_9 &= \omega.\omega.\alpha.\beta.\omega
\end{aligned} \tag{1.30}$$

where for the sake of simplicity I've replaced the symbol $\otimes$ with a dot. These 10 elements of $\mathbf{R}(32)$ satisfy

$$\frac{1}{2}(\Upsilon_k\Upsilon_l + \Upsilon_l\Upsilon_k) = \eta_{kl}\epsilon_{32}, \tag{1.31}$$

where $\{\eta_{kl}\} = \text{diag}(+---------)$, and ϵ_{32} is the 32×32 identity matrix. Furthermore, the pseudoscalar

$$\Upsilon_0\Upsilon_1...\Upsilon_9 = \pm\alpha.\epsilon.\epsilon.\epsilon.\epsilon = \pm\begin{bmatrix} \epsilon_{16} & 0 \\ 0 & -\epsilon_{16} \end{bmatrix} \tag{1.32}$$

(I leave it to the reader to work out the sign). Clearly then the Υ_k, k=0,...,9, are 1-vector generators of the Clifford algebra $\mathbf{R}_{1,9}$. (Note: $\mathbf{R}_{9,1}$ is easily constructable by changing the lead β in Υ_0 to ω, and the lead ω's in Υ_k, k=1,...,9, to β's.)

Note that in this case the chiral projectors

$$\frac{1}{2}(\epsilon.\epsilon.\epsilon.\epsilon.\epsilon \pm \alpha.\epsilon.\epsilon.\epsilon.\epsilon) = \begin{bmatrix} \epsilon_{16} & 0 \\ 0 & 0 \end{bmatrix} \text{ and } \begin{bmatrix} 0 & 0 \\ 0 & \epsilon_{16} \end{bmatrix}, \tag{1.33}$$

are elements of the Clifford algebra $\mathbf{R}_{1,9}$. Therefore it is not necessary to complexify the Clifford algebra and its spinors to project Weyl spinors. So $\mathbf{R}_{1,9}$ spinors can be simultaneously Majorana and Weyl. In the present representation a Majorana spinor would be a 32×1 real column, and a Majorana/Weyl spinor would be all zero in the first or last 16 positions.

1.3. Conjugations and Spinors.

There is a pair of natural (product reversing) conjugations on any Clifford algebra [1] determined by their action on the 1-vector generators:

$$\boxed{\Upsilon_\alpha^\wedge = \Upsilon_\alpha} \tag{1.34}$$

and

$$\boxed{\Upsilon_\alpha^\vee = -\Upsilon_\alpha.} \tag{1.35}$$

In both cases, because these maps are antiautomorphic,

$$\boxed{\Upsilon_{\alpha\beta}^\wedge = \Upsilon_{\alpha\beta}^\vee = -\Upsilon_{\alpha\beta}.} \tag{1.36}$$

In general,

$$\begin{array}{lll}
1-\textit{vector} & x^\wedge = +x & x^\vee = -x \\
2-\textit{vector} & x^\wedge = -x & x^\vee = -x \\
3-\textit{vector} & x^\wedge = -x & x^\vee = +x \\
4-\textit{vector} & x^\wedge = +x & x^\vee = +x \\
 & & \\
5-\textit{vector} & x^\wedge = +x & x^\vee = -x \\
6-\textit{vector} & x^\wedge = -x & x^\vee = -x \\
7-\textit{vector} & x^\wedge = -x & x^\vee = +x \\
8-\textit{vector} & x^\wedge = +x & x^\vee = +x \\
 & & \\
9-\textit{vector} & x^\wedge = +x & x^\vee = -x \\
10-\textit{vector} & x^\wedge = -x & x^\vee = -x \\
11-\textit{vector} & x^\wedge = -x & x^\vee = +x \\
12-\textit{vector} & x^\wedge = +x & x^\vee = +x
\end{array} \tag{1.37}$$

In certain cases these conjugations correspond to ordinary Hermitian conjugation, denoted $x^\dagger$:

$$x^\wedge = x^\dagger : \quad x \in \mathbf{R}_{n,0}, \quad n = 1,2,3,...;$$
$$x^\vee = x^\dagger : \quad x \in \mathbf{R}_{0,n}, \quad n = 1,2,3,.... \tag{1.38}$$

In all other cases the conjugations are more difficult to define.

In the representations 1.19 of $\mathbf{R}_{3,1}$, and 1.23,26 of $\mathbf{R}_{1,3}$, these conjugations are defined the same for each. Given $x, y, u, v \in \mathbf{C}(2)$ (for the 1.19 and 1.26 cases), or (equivalently) $x, y, u, v \in \mathbf{C} \otimes \mathbf{Q}$ (for the 1.23 case), we define

$$\begin{bmatrix} x & u \\ v & y \end{bmatrix}^\wedge = \begin{bmatrix} y^\dagger & u^\dagger \\ v^\dagger & x^\dagger \end{bmatrix}, \tag{1.39}$$

and

$$\begin{bmatrix} x & u \\ v & y \end{bmatrix}^\vee = \begin{bmatrix} y^\dagger & -u^\dagger \\ -v^\dagger & x^\dagger \end{bmatrix}. \tag{1.40}$$

These conjugations are generalizable to within a sign or phase to the spinors of the complexifications of $\mathbf{R}_{3,1}$, and $\mathbf{R}_{1,3}$. In particular, let

$$\psi_l,\ \psi_r \in \mathbf{C}^2,$$

and

$$\Psi = \begin{bmatrix} \psi_l \\ \psi_r \end{bmatrix} \in \mathbf{C}^4,$$

a Dirac spinor for the Dirac algebra, $\mathbf{C} \otimes \mathbf{R}_{1,3}$, as represented in 1.26. Then we define

$$\Psi^\wedge = \begin{bmatrix} \psi_l \\ \psi_r \end{bmatrix}^\wedge = [\ \psi_r^\dagger \quad \psi_l^\dagger\ ], \quad (\Psi^\wedge)^\wedge = \Psi, \tag{1.41}$$

and

$$\Psi^\vee = \begin{bmatrix} \psi_l \\ \psi_r \end{bmatrix}^\vee = [\ \psi_r^\dagger \quad -\psi_l^\dagger\ ], \quad (\Psi^\vee)^\vee = -\Psi. \tag{1.42}$$

The former I shall call the spinor conjugate associated with the Clifford algebra. Note that in both cases,

$$(\Psi^\wedge \gamma_\mu \Psi)^\wedge = \Psi^\wedge \gamma_\mu^\wedge (\Psi^\wedge)^\wedge = \Psi^\wedge \gamma_\mu \Psi, \tag{1.43}$$

$$(\Psi^\vee\gamma_\mu\Psi)^\vee = \Psi^\vee\gamma_\mu^\vee(\Psi^\vee)^\vee = \Psi^\vee(-\gamma_\mu)(-\Psi) = \Psi^\vee\gamma_\mu\Psi. \quad (1.44)$$

Also note that in the Dirac algebra case $\Psi^\wedge$ is usually denoted $\bar{\Psi}$, and

$$\boxed{\Psi^\wedge = \bar{\Psi} = \Psi^\dagger\gamma_0.} \quad (1.45)$$

NILPOTENT CLIFFORD ALGEBRAS

The reader is warned that what follows is an unconventional approach to Clifford algebra theory and to the best of my knowledge it has not found its way into the mainstream literature on physical model building **[2][3][4]**. It will also play no role in what follows, although the attentive reader will see how it might.

Define

$$\boxed{\mathbf{N}_{p,q} \simeq \mathbf{R}_{p+1,q} \simeq \mathbf{R}_{q+1,p} \simeq \mathbf{N}_{q,p}.} \quad (1.46)$$

This is what I refer to as the nilpotent Clifford algebra of the pseudo-orthogonal space $\mathbf{R}^{p,q}$. Note that unlike the ordinary Clifford algebra case, where $\mathbf{R}_{p,q}$ and $\mathbf{R}_{q,p}$ are seldom isomorphic, $\mathbf{N}_{p,q}$ is always isomorphic to $\mathbf{N}_{q,p}$. For example,

$$\mathbf{N}_{1,3} \simeq \mathbf{N}_{3,1} \simeq \mathbf{C}(4), \quad (1.47)$$

the Dirac algebra for (1,3)-spacetime.

Let

$$\Upsilon,\ \Upsilon_1, ..., \Upsilon_n \quad (1.48)$$

be a 1-vector basis for $\mathbf{R}_{p+1,q}$, where $n = p + q$. The Υ_α, $\alpha = 1, ..., n$ are assumed to satisfy 1.1, while

$$\Upsilon^2 = 1,$$
$$\Upsilon\Upsilon_\alpha + \Upsilon_\alpha\Upsilon = 0. \quad (1.49)$$

(Note that the collection of 1-vectors 1.48, given the commutator product, generates the conformal Lie algebra $so(p+1, q+1)$.)

Define

$$\boxed{\mu_\pm = \tfrac{1}{2}(1 \pm \Upsilon),} \quad (1.50)$$

which are orthogonal projection operators ($\mu_\pm\mu_\mp = 0$), and define

$$\boxed{\Upsilon_\alpha^\pm = \mu_\pm\Upsilon_\alpha.} \quad (1.51)$$

Because of 1.49 these elements are nilpotent. In fact, for all $\alpha, \beta \in \{1, ..., n\}$,

$$\Upsilon_\alpha^\pm \Upsilon_\beta^\pm = \mu_\pm \Upsilon_\alpha \mu_\pm \Upsilon_\beta = \mu_\pm \mu_\mp \Upsilon_\alpha \Upsilon_\beta = 0.$$

(In the (1,3)-case, $\mu_\pm$ are the chiral projectors, so nilpotent Clifford algebras have chirality built in.) Also, if $\alpha, \beta, \nu \in \{1, ..., n\}$, then

$$\begin{aligned} \Upsilon_\alpha^\pm \Upsilon_\nu^\mp \Upsilon_\beta^\pm + \Upsilon_\beta^\pm \Upsilon_\nu^\mp \Upsilon_\alpha^\pm &= \mu_\pm(\Upsilon_\alpha \Upsilon_\nu \Upsilon_\beta + \Upsilon_\beta \Upsilon_\nu \Upsilon_\alpha) \\ &= 2\Upsilon_\alpha^\pm \eta_{\nu\beta} + 2\Upsilon_\beta^\pm \eta_{\nu\alpha} - 2\Upsilon_\nu^\pm \eta_{\alpha\beta}. \end{aligned}$$

We shall write these results in the following ways:

$$\boxed{\Upsilon_\alpha^\pm \Upsilon_\beta^\pm = 0,} \tag{1.52}$$

and

$$\boxed{\begin{aligned} \Upsilon_\alpha^+ \Upsilon_\nu^- \Upsilon_\beta^+ + \Upsilon_\beta^+ \Upsilon_\nu^- \Upsilon_\alpha^+ &= 2\eta_{\alpha\nu} \Upsilon_\beta^+ + 2\eta_{\beta\nu} \Upsilon_\alpha^+ - (\eta_{\alpha\beta} + \eta_{\beta\alpha}) \Upsilon_\nu^+ \\ \Upsilon_\alpha^- \Upsilon_\nu^+ \Upsilon_\beta^- + \Upsilon_\beta^- \Upsilon_\nu^+ \Upsilon_\alpha^- &= 2\Upsilon_\alpha^- \eta_{\nu\beta} + 2\Upsilon_\beta^- \eta_{\nu\alpha} - (\eta_{\alpha\beta} + \eta_{\beta\alpha}) \Upsilon_\nu^-. \end{aligned}} \tag{1.53}$$

SYMPLECTIC NILPOTENT CLIFFORD ALGEBRA

The equations 1.52 and 1.53 are sufficient to completely determine $\mathbf{N}_{p,q}$. They are presented in what must seem a peculiar way to allow for a generalization to the symplectic (or alternating) case, as opposed to the pseudo-orthogonal.

In particular, replace $\Upsilon_\alpha^\pm$ by $S_i^\pm$, $i = 1, 2$, so $n = 2$, and replace $\eta_{\alpha\beta}$ by ϵ_{ij}, such that

$$\epsilon_{12} = -\epsilon_{21} = 1; \ \epsilon_{11} = \epsilon_{22} = 0.$$

In this case the equations 1.52,53 reduce to

$$\boxed{S_i^\pm S_j^\pm = 0,} \tag{1.54}$$

and

$$\boxed{\begin{aligned} S_i^+ S_j^- S_k^+ + S_k^+ S_j^- S_i^+ &= 2\epsilon_{ij} S_k^+ + 2\epsilon_{kj} S_i^+, \\ S_i^- S_j^+ S_k^- + S_k^- S_j^+ S_i^- &= 2S_i^- \epsilon_{jk} + 2S_k^- \epsilon_{ji}. \end{aligned}} \tag{1.55}$$

Therefore,

$$
\begin{aligned}
S_1^+ S_2^- S_1^+ &= 2S_1^+; \quad S_1^- S_2^+ S_1^- = -2S_1^-;\\
S_2^+ S_1^- S_2^+ &= -2S_2^+; \quad S_2^- S_1^+ S_2^- = 2S_2^-;\\
S_1^+ S_1^- S_2^+ + S_2^+ S_1^- S_1^+ &= -2S_1^+; \quad S_1^- S_1^+ S_2^- + S_2^- S_1^+ S_1^- = 2S_1^-;\\
S_1^+ S_2^- S_2^+ + S_2^+ S_2^- S_1^+ &= 2S_2^+; \quad S_1^- S_2^+ S_2^- + S_2^- S_2^+ S_1^- = -2S_2^-;\\
S_1^\pm S_1^\mp S_1^\pm &= 0; \quad S_2^\pm S_2^\mp S_2^\pm = 0.
\end{aligned}
\tag{1.56}
$$

Using the equations 1.54,56 one can show that the algebra generated by the $S_i^\pm$ is isomorphic to the 18-dimensional matrix algebra ${}^2\mathbf{R}(3)$. We can represent the generators as

$$
S_1^+ = \begin{bmatrix} \boxed{\begin{matrix} \cdot & \cdot & 1 \\ \cdot & \cdot & 1 \\ \cdot & \cdot & \cdot \end{matrix}} & \\ & \boxed{\begin{matrix} \cdot & \cdot & \cdot \\ \cdot & \cdot & \cdot \\ -1 & 1 & \cdot \end{matrix}} \end{bmatrix}, \quad
S_2^+ = \begin{bmatrix} \boxed{\begin{matrix} \cdot & \cdot & 1 \\ \cdot & \cdot & -1 \\ \cdot & \cdot & \cdot \end{matrix}} & \\ & \boxed{\begin{matrix} \cdot & \cdot & \cdot \\ \cdot & \cdot & \cdot \\ 1 & 1 & \cdot \end{matrix}} \end{bmatrix},
$$

$$
S_1^- = \begin{bmatrix} \boxed{\begin{matrix} \cdot & \cdot & \cdot \\ \cdot & \cdot & \cdot \\ -1 & 1 & \cdot \end{matrix}} & \\ & \boxed{\begin{matrix} \cdot & \cdot & -1 \\ \cdot & \cdot & -1 \\ \cdot & \cdot & \cdot \end{matrix}} \end{bmatrix}, \quad
S_2^- = \begin{bmatrix} \boxed{\begin{matrix} \cdot & \cdot & \cdot \\ \cdot & \cdot & \cdot \\ 1 & 1 & \cdot \end{matrix}} & \\ & \boxed{\begin{matrix} \cdot & \cdot & -1 \\ \cdot & \cdot & 1 \\ \cdot & \cdot & \cdot \end{matrix}} \end{bmatrix}.
\tag{1.57}
$$

In the pseudo-orthogonal case, the elements $\Upsilon_\alpha = \Upsilon_\alpha^+ + \Upsilon_\alpha^-$ are ordinary Clifford algebra 1-vectors, and the set of all commutators

$$\Upsilon_{\alpha\beta} = \frac{1}{2}[\Upsilon_\alpha, \Upsilon_\beta]$$

form a basis for $so(p,q)$, which generates the invariance group of the bilinear form associated with $\eta_{\alpha\beta}$.

In the symplectic case we define

$$\boxed{S_i = S_i^+ + S_i^-,\ i = 1, 2,} \tag{1.58}$$

and we replace commutators with anticommutators:

$$S_{11} = \frac{1}{2}\{S_1, S_1\} = \begin{bmatrix} \boxed{\begin{matrix} -1 & 1 & . \\ -1 & 1 & . \\ . & . & . \end{matrix}} & \\ & \boxed{\begin{matrix} 1 & -1 & . \\ 1 & -1 & . \\ . & . & . \end{matrix}} \end{bmatrix},$$

$$S_{22} = \frac{1}{2}\{S_1, S_1\} = \begin{bmatrix} \boxed{\begin{matrix} 1 & 1 & . \\ -1 & -1 & . \\ . & . & . \end{matrix}} & \\ & \boxed{\begin{matrix} -1 & -1 & . \\ 1 & 1 & . \\ . & . & . \end{matrix}} \end{bmatrix},$$

$$S_{12} = S_{21} = \frac{1}{2}\{S_1, S_2\} = \begin{bmatrix} \boxed{\begin{matrix} 0 & 1 & . \\ 1 & 0 & . \\ . & . & . \end{matrix}} & \\ & \boxed{\begin{matrix} 0 & -1 & . \\ -1 & 0 & . \\ . & . & . \end{matrix}} \end{bmatrix}$$

(no entry, or a dot entry, signifies a zero component). These matrices constitute a basis for $sp(2)$, which generates the invariance group of the alternating bilinear form associated with ϵ_{ij}. Together with the S_i, $i = 1, 2$, and given the appropriate mix of commutator and anticommutator products, these five elements constitute a basis for the Lie superalgebra $osp(1,2)$ [5].

Clearly this approach to the idea of geometric algebra is richer than the conventional one. To this point, however, little has been done to apply these symplectic nilpotent Clifford algebras, nor is much known about their general mathematical properties. They are presented here just as an idea.

A final word about Clifford algebras in general: they are, like Lie groups, profligate. There are too many of them, an infinite number of both Lie groups and Clifford algebras that are physically irrelevant, not a part of the design of reality. This is and always was the problem with GUT theories based on a unifying large Lie group, and it is and always will be the problem with unification theories based on large Clifford algebras. In both cases it is the principal of the educated guess that leads to the choice of unifying algebraic object. This is unsatisfactory. Nature can not be so arbitrarily ugly. The first three division algebras, **R**, **C**, and **Q**, are clearly intimately linked to Clifford algebras in general, and it will be demonstrated in chapter 2 that the octonions, despite their nonassociativity, give rise to Clifford algebras as well (the algebra of left adjoint actions of **O** on **O** is isomorphic to the associative algebra **R**(8), so it could be incorporated into the Clifford algebra tables listed in this section (for that matter, in the isomorphisms listed that involve **Q** it should be specified that it is **Q**'s left adjoint algebra that is being used, for it is this that acts on the column spinor space)). As will be pointed out in chapter 6 the sequence of division algebras **R**, **C**, **Q**, and **O** is linked to the sequence of Clifford algebras of spacetimes of dimension 3,4,6, and 10, critical dimensions for classical string theory. This monograph, however, is not about string theory, and it is not this link of the division algebras to the critical dimensions of string theory that gave birth to my interest in these algebras. Instead it was the isomorphism $\mathbf{P} \simeq \mathbf{C} \otimes \mathbf{Q}$, linking the Pauli algebra of 4-dimensional spacetime to the tensor product of the first two hypercomplex division algebras, and the fact that the use of $\mathbf{C} \otimes \mathbf{Q}$ instead of **P** in building the Dirac algebra leads automatically to the consideration of a $U(1) \times SU(2)$ internal symmetry. Tensoring $\mathbf{C} \otimes \mathbf{Q}$ with **O** connects to 10-dimensional spacetime in the same way, and leads to the inclusion of an $SU(3)$ internal symmetry and the reduction of 10-dimensional spacetime to 4-dimensional **[6]**. In my opinion these are the cleanest and clearest links of the division algebras to the design of physical reality. This is not to say that the string theory links **[7][8]**, or, for that matter, the quantum mechanical links via Jordan algebras **[9]**, are wrong or irrelevant, just that their ultimate relevance has not been demonstrated. It is my personal hope, in that these different approaches to application are not inherently incompatible, that eventually they may combine, and their union be more universally unifying than any is individually.

1.4. Algebraic Fundamentals of the Standard Model.

Our observable spacetime has 1 time and 3 space dimensions. However, instead of housing physical fermion fields in the spinor spaces of either the Clifford algebra $\mathbf{R}_{1,3}$ or $\mathbf{R}_{3,1}$, Nature requires a complexification of the Clifford algebra ,

$$\mathbf{C} \otimes \mathbf{R}_{1,3} \simeq \mathbf{C} \otimes \mathbf{R}_{3,1} \simeq \mathbf{C}(4),$$

and She exploits Dirac spinor fields involving 4 complex components. However, these are not the fundamental spinors of physics, for it is observed that the two Weyl spinors of which a Dirac spinor is composed behave differently in the real world.

Each fundamental fermion (leptons and quarks) is representable by a pair of complex 2 component Weyl spinor fields (left and right Chiral projections). [Righthanded neutrino fields are thought by some not to exist in Nature, as they evidently can have no interactions save for gravitational. However, no evidence for their absence exists.]

The internal gauge symmetry from which vector boson gauge fields arise has been discovered to be

$$U(1) \times SU(2) \times SU(3),$$

the so-called standard symmetry. Lepton and quark fields interact with each other via these $1 + 3 + 8 = 12$ gauge fields. The nature of each fermion field is determined by its $U(1) \times SU(2) \times SU(3)$ charges, which in turn determine how the fermions will interact with the gauge bosons of the theory.

The standard symmetry is put into the theory by hand, as are the various leptoquark charges.

Lepton and quark fields are observed to fall into either $SU(2)$ singlets or $SU(2)$ doublets. Lefthanded Weyl spinor fields fall into doublets; righthanded Weyl spinor fields fall into singlets. Hence $SU(2)$ is said to be chiral; it violates parity.

Leptons fall into $SU(3)$ singlets, and quarks into $SU(3)$ triplets. This is independent of handedness, so $SU(3)$ is said to be nonchiral; it conserves parity.

The leptoquark multiplet structure associated with $SU(2)$ and $SU(3)$ is put into the theory by hand, as is the chiral nature of $SU(2)$.

An exact symmetry will mix the fields of a multiplet while preserving masses. Not all of the groups of the standard symmetry are observed to be exact. $SU(3)$ is exact (hence its gauge fields, the gluons, are massless), as is a $U(1) \subset U(1) \times SU(2)$ giving rise to the massless photon, the gauge field of electromagnetism. The rest of $U(1) \times SU(2)$ is broken, and the corresponding gauge fields ($W^{\pm}$, Z^0) are massive. Scalar fields are introduced that are designed to break most of $U(1) \times SU(2)$, but leave a $U(1) \times SU(3)$ symmetry exact. This process, called spontaneous symmetry breaking, accounts for the mass differences observed between the charge (-1) and charge (0) leptons, and between the charge $(-1/3)$ and $(2/3)$ quarks. Because $SU(3)$ is exact, the three "colors" of quarks in an $SU(3)$ triplet have the same mass.

Finally, this pattern of fermion multiplets, which accounts at lowest level for the electron and its neutrino, the up-quark and down-quark (3 colors each; a total of $2(1+3) = 8$ fermion fields called a family), is replicated at least twice more. Each of these additional families is distinguished from the first, and from each other, only in the masses of its fields. For example, each of the masses of the particles in the second family, the muon and its neutrino, the strange-quark and charm-quark, is more massive than the corresponding fields of the electron family.

The explication of these algebraic features via the mathematics of the division algebras is a central goal of the first half of this monograph.

2. Division Algebras Alone.

2.1. Mostly Octonions.

The octonion algebra, **O**, is generally developed as an extension of the quaternion algebra, **Q**. Let q_i, i=1,2,3, be a conventional basis for the hypercomplex quaternions. These elements associate, anticommute, and satisfy $q_i^2 = -1$. Together with the cyclic multiplication rule

$$\boxed{q_i q_{i+1} = q_{i+2},} \tag{2.1}$$

i=1,...,3, indices modulo 3, from 1 to 3, this completely determines the multiplication table for **Q**.

$$\begin{bmatrix} 1 & q_1 & q_2 & q_3 \\ q_1 & -1 & q_3 & -q_2 \\ q_2 & -q_3 & -1 & q_1 \\ q_3 & q_2 & -q_1 & -1 \end{bmatrix}.$$

Relabel these quaternion units e_i, i=1,2,3, and introduce a new unit, e_7, anticommuting with each of the e_i, which satisfies $e_7^2 = -1$. Define three more units:

$$e_4 = e_1 e_7,\ e_5 = e_2 e_7,\ e_6 = e_3 e_7. \tag{2.2}$$

Let **O** be the real algebra (with identity) generated from the e_a a=1,...,7, such that

$$\{q_1 \rightarrow e_a,\ q_2 \rightarrow e_b,\ q_3 \rightarrow e_c\}$$

defines an injection of **Q** into **O** for

$$(a,b,c) = (1,2,3),(1,7,4),(2,7,5),(3,7,6),(1,6,5),(2,4,6),(3,5,4).$$

Therefore, for example,

$$e_1(e_7 e_5) = e_1 e_2 = e_3 = -(-e_3) = -e_4 e_5 = -(e_1 e_7) e_5.$$

So unlike the complexes and quaternions, the octonions are nonassociative. In general, if e_a, e_b, and e_c generate a quaternion subalgebra of **O**, then they associate, and if not, then they do not.

Like **C** and **Q**, however, **O** is a division algebra, and it is normed. In particular, if $x = x^0 + x^a e_a$, (sum a=1,...,7), and

$$\boxed{x^\dagger = x^0 - x^a e_a}$$

(an antiautomorphism), then

$$\boxed{\|x\|^2 = x^\dagger x = \sum_{a=0}^{7} x^a x^a} \tag{2.3}$$

defines the square of the norm of x (so

$$\boxed{x^{-1} = x^\dagger / \|x\|^2).}$$

This octonion multiplication is not, however, the most natural (see section 2.6), and it will not be employed here. Again let $e_a, a = 1, ..., 7$, represent the hypercomplex units of **O**, but now adopt the cyclic multiplication rule

$$\boxed{e_a e_{a+1} = e_{a+5} = e_{a-2},} \tag{2.4}$$

a=1,...,7, all indices modulo 7, from 1 to 7 . In particular,

$$\{q_1 \to e_a, q_2 \to e_{a+1}, q_3 \to e_{a+5}\}$$

define injections of **Q** into **O** for a=1,...,7. I am accustomed to using the symbol e_0 to represent unity, and I bother to remember that although $7 = 0$ mod 7, $e_7 \neq e_0$, and in the multiplication rule 2.4 the indices range from 1 to 7, and the index 0 is not subject to the rule. (In **[10]** ∞ is used as the index for unity, and this has advantages, which I find intermittently persuasive.)

[Note: Another cyclic multiplication rule for **O**, dual to that above in a way outlined in section 2.6, is

$$e_a e_{a+1} = e_{a+3} = e_{a-4}. \tag{2.5}$$

This one is more commonly used than 2.4. Indeed, of the two cyclic multiplication rules 2.4 and 2.5 I may be the only one who uses 2.4. None of the mathematical or theoretical development of this monograph depends on the choice, and both are pretty in the same way. Using 2.5 the three octonion units $\{e_1, e_2, e_4\}$ (all subscripts powers of 2) associate:

$$e_1(e_2e_4) = (e_1e_2)e_4 = -1.$$

In this case they generate a **Q** subalgebra. Using the rule 2.4 these same units do not associate:

$$e_1(e_2e_4) = -(e_1e_2)e_4 = e_7.$$

In this case they generate the whole of **O**. I thought that a nice property, and on that questionable basis I made my choice.]

Both the octonion multiplication rules 2.4 and 2.5 have some very nice properties. For example,

$$\boxed{\text{if } e_ae_b = e_c, \text{ then } e_{(2a)}e_{(2b)} = e_{(2c)}.} \tag{2.6}$$

2.6 in combination with 2.4 immediately implies

$$\begin{aligned} e_ae_{a+2} &= e_{a+3}, \\ e_ae_{a+4} &= e_{a+6} \end{aligned} \tag{2.7}$$

(so

$$e_ae_{a+2^n} = e_{a-2^{n+1}},$$

or

$$e_ae_{a+b} = [b^3 \bmod 7]e_{a-2b^4},\ b = 1, ..., 6,$$

where b^3 out front provides the sign of the product (modulo 7, $1^3 = 2^3 = 4^3 = 1$, and $3^3 = 5^3 = 6^3 = -1$)). Also, 2(7)=7 mod 7, so 2.6 and 2.4 imply

$$e_7e_1 = e_5,\quad e_7e_2 = e_3,\quad e_7e_4 = e_6. \tag{2.8}$$

These modulo 7 periodicity properties are reflected in the full multiplication table:

$$\begin{bmatrix} 1 & e_1 & e_2 & e_3 & e_4 & e_5 & e_6 & e_7 \\ e_1 & -1 & e_6 & e_4 & -e_3 & e_7 & -e_2 & -e_5 \\ e_2 & -e_6 & -1 & e_7 & e_5 & -e_4 & e_1 & -e_3 \\ e_3 & -e_4 & -e_7 & -1 & e_1 & e_6 & -e_5 & e_2 \\ e_4 & e_3 & -e_5 & -e_1 & -1 & e_2 & e_7 & -e_6 \\ e_5 & -e_7 & e_4 & -e_6 & -e_2 & -1 & e_3 & e_1 \\ e_6 & e_2 & -e_1 & e_5 & -e_7 & -e_3 & -1 & e_4 \\ e_7 & e_5 & e_3 & -e_2 & e_6 & -e_1 & -e_4 & -1 \end{bmatrix}. \tag{2.9}$$

The naturalness of this table is reflected in the matrix of signs derived from 2.9:

$$O = \begin{bmatrix} 1 & 1 & 1 & 1 & 1 & 1 & 1 & 1 \\ 1 & -1 & 1 & 1 & -1 & 1 & -1 & -1 \\ 1 & -1 & -1 & 1 & 1 & -1 & 1 & -1 \\ 1 & -1 & -1 & -1 & 1 & 1 & -1 & 1 \\ 1 & 1 & -1 & -1 & -1 & 1 & 1 & -1 \\ 1 & -1 & 1 & -1 & -1 & -1 & 1 & 1 \\ 1 & 1 & -1 & 1 & -1 & -1 & -1 & 1 \\ 1 & 1 & 1 & -1 & 1 & -1 & -1 & -1 \end{bmatrix}. \qquad (2.10)$$

[This is what is called a normalized Hadamard matrix of order 8 **[10]**. It is normalized because the first row and column are all 1's, and it is a Hadamard matrix in containing only 1's and -1's, and in satisfying $OO^\dagger = 8I$, where $O^\dagger$ is the transpose of O, and I is the 8x8 identity matrix.]

Note that if $a \neq 0$, $b \neq 0$, then the components $O_{a,b} = O_{a+1,b+1}$, indices from 1 to 7, modulo 7 (first row and column of O are given the index 0). Let O_a be the ath row of O, a=0,1,...,7, and define the product

$$O_a \bullet O_b = O_{a,b} O_c, \qquad (2.11)$$

where the components of O_c are

$$O_{c,d} = O_{a,d} O_{b,d},$$

for each d=0,1,...,7, and $O_{a,b}$ gives a sign to the product. For example,

$$O_1 \bullet O_2$$

$$= +[1 \cdot 1, (-1)\cdot(-1), 1\cdot(-1), 1\cdot 1, (-1)\cdot 1, 1\cdot(-1), (-1)\cdot 1, (-1)\cdot(-1)]$$

$$= +[1, 1, -1, 1, -1, -1, -1, 1] = O_6,$$

where the plus sign out front arises from the component $O_{1,2} = +1$. The resulting multiplication table of the $O_a' s$ is exactly the same as 2.9, giving rise to the obvious isomorphism $e_a \rightarrow O_a, a = 0, 1, ..., 7$.

The quaternion algebra arises in exactly the same way from the Hadamard matrix of order 4:

$$Q = \begin{bmatrix} 1 & 1 & 1 & 1 \\ 1 & -1 & 1 & -1 \\ 1 & -1 & -1 & 1 \\ 1 & 1 & -1 & -1 \end{bmatrix}. \qquad (2.12)$$

In this case, while index cycling (modulo 3) is an automorphism, index doubling is a (product reversing) antiautomorphism.

Likewise the complexes arise from the order 2 Hadamard matrix

$$C = \begin{bmatrix} 1 & 1 \\ 1 & -1 \end{bmatrix}. \tag{2.13}$$

Index cycling and doubling, both modulo 1, have no effect here.

2.2. Adjoint algebras.

The octonion algebra is nonassociative and so is not representable as a matrix algebra. The adjoint algebras of left and right actions of **O** on itself are associative. For example, let $u_1, ..., u_n, x$ be elements of **O**. Consider the left adjoint map

$$x \rightarrow u_n(...(u_2(u_1 x))...). \tag{2.14}$$

The nesting of parentheses forces the products to occur in a certain order, hence this algebra of left-actions is trivially associative, and it is representable by a matrix algebra. One such representation can be derived immediately from the multiplication table 2.9. For example, the actions

$$x \rightarrow e_1 x \equiv e_{L1}[x], \text{ and } x \rightarrow x e_1 \equiv e_{R1}[x] \tag{2.15}$$

can be identified with the matrices

$$e_{L1} \rightarrow \begin{bmatrix} \cdot & -1 & \cdot & \cdot & \cdot & \cdot & \cdot & \cdot \\ 1 & \cdot & \cdot & \cdot & \cdot & \cdot & \cdot & \cdot \\ \cdot & \cdot & \cdot & \cdot & \cdot & \cdot & -1 & \cdot \\ \cdot & \cdot & \cdot & \cdot & -1 & \cdot & \cdot & \cdot \\ \cdot & \cdot & \cdot & 1 & \cdot & \cdot & \cdot & \cdot \\ \cdot & \cdot & \cdot & \cdot & \cdot & \cdot & \cdot & -1 \\ \cdot & \cdot & 1 & \cdot & \cdot & \cdot & \cdot & \cdot \\ \cdot & \cdot & \cdot & \cdot & \cdot & 1 & \cdot & \cdot \end{bmatrix},$$

$$e_{R1} \rightarrow \begin{bmatrix} \cdot & -1 & \cdot & \cdot & \cdot & \cdot & \cdot & \cdot \\ 1 & \cdot & \cdot & \cdot & \cdot & \cdot & \cdot & \cdot \\ \cdot & \cdot & \cdot & \cdot & \cdot & \cdot & 1 & \cdot \\ \cdot & \cdot & \cdot & \cdot & 1 & \cdot & \cdot & \cdot \\ \cdot & \cdot & \cdot & -1 & \cdot & \cdot & \cdot & \cdot \\ \cdot & \cdot & \cdot & \cdot & \cdot & \cdot & \cdot & 1 \\ \cdot & \cdot & -1 & \cdot & \cdot & \cdot & \cdot & \cdot \\ \cdot & \cdot & \cdot & \cdot & \cdot & -1 & \cdot & \cdot \end{bmatrix} \tag{2.16}$$

(only nonzero entries are indicated). Note that

$$e_{L1}[x] = e_1(x^0e_0 + x^1e_1 + x^2e_2 + x^3e_3 + x^4e_4 + x^5e_5 + x^6e_6 + x^7e_7)$$

$$= e_1[\ e_0 \quad e_1 \quad e_2 \quad e_3 \quad e_4 \quad e_5 \quad e_6 \quad e_7\ ]\begin{bmatrix} x^0 \\ x^1 \\ x^2 \\ x^3 \\ x^4 \\ x^5 \\ x^6 \\ x^7 \end{bmatrix}$$

$$= [\ e_1 \quad -e_0 \quad e_6 \quad e_3 \quad -e_3 \quad e_7 \quad -e_2 \quad -e_5\ ]\begin{bmatrix} x^0 \\ x^1 \\ x^2 \\ x^3 \\ x^4 \\ x^5 \\ x^6 \\ x^7 \end{bmatrix}$$

$$= [\ e_0 \quad e_1 \quad e_2 \quad e_3 \quad e_4 \quad e_5 \quad e_6 \quad e_7\ ]\begin{bmatrix} . & -1 & . & . & . & . & . & . \\ 1 & . & . & . & . & . & . & . \\ . & . & . & . & . & . & -1 & . \\ . & . & . & . & -1 & . & . & . \\ . & . & . & 1 & . & . & . & . \\ . & . & . & . & . & . & . & -1 \\ . & . & 1 & . & . & . & . & . \\ . & . & . & . & . & 1 & . & . \end{bmatrix}\begin{bmatrix} x^0 \\ x^1 \\ x^2 \\ x^3 \\ x^4 \\ x^5 \\ x^6 \\ x^7 \end{bmatrix}$$

which gives rise to the first of the matrix identifications in 2.16 (please note that we can identify these adjoint elements with matrices because of the associativity of $\mathbf{O}_L$, the algebra of left actions of $\mathbf{O}$ on itself; this is not possible for the nonassociative $\mathbf{O}$).

Note that $e_2e_6 = e_3e_4 = e_5e_7 = e_1$, but because of the nonassociativity of $\mathbf{O}$, for example,

$$e_1x = (e_2e_6)x \neq e_2(e_6x) \equiv e_{L26}[x] \tag{2.17}$$

in general. In particular,

$$e_{L26} \to \begin{bmatrix} \cdot & -1 & \cdot & \cdot & \cdot & \cdot & \cdot & \cdot \\ 1 & \cdot & \cdot & \cdot & \cdot & \cdot & \cdot & \cdot \\ \cdot & \cdot & \cdot & \cdot & \cdot & \cdot & -1 & \cdot \\ \cdot & \cdot & \cdot & \cdot & 1 & \cdot & \cdot & \cdot \\ \cdot & \cdot & \cdot & -1 & \cdot & \cdot & \cdot & \cdot \\ \cdot & \cdot & \cdot & \cdot & \cdot & \cdot & \cdot & 1 \\ \cdot & \cdot & 1 & \cdot & \cdot & \cdot & \cdot & \cdot \\ \cdot & \cdot & \cdot & \cdot & \cdot & -1 & \cdot & \cdot \end{bmatrix},$$

$$e_{L34} \to \begin{bmatrix} \cdot & -1 & \cdot & \cdot & \cdot & \cdot & \cdot & \cdot \\ 1 & \cdot & \cdot & \cdot & \cdot & \cdot & \cdot & \cdot \\ \cdot & \cdot & \cdot & \cdot & \cdot & \cdot & 1 & \cdot \\ \cdot & \cdot & \cdot & \cdot & -1 & \cdot & \cdot & \cdot \\ \cdot & \cdot & \cdot & 1 & \cdot & \cdot & \cdot & \cdot \\ \cdot & \cdot & \cdot & \cdot & \cdot & \cdot & \cdot & 1 \\ \cdot & \cdot & -1 & \cdot & \cdot & \cdot & \cdot & \cdot \\ \cdot & \cdot & \cdot & \cdot & \cdot & -1 & \cdot & \cdot \end{bmatrix},$$

$$e_{L57} \to \begin{bmatrix} \cdot & -1 & \cdot & \cdot & \cdot & \cdot & \cdot & \cdot \\ 1 & \cdot & \cdot & \cdot & \cdot & \cdot & \cdot & \cdot \\ \cdot & \cdot & \cdot & \cdot & \cdot & \cdot & 1 & \cdot \\ \cdot & \cdot & \cdot & \cdot & 1 & \cdot & \cdot & \cdot \\ \cdot & \cdot & \cdot & -1 & \cdot & \cdot & \cdot & \cdot \\ \cdot & \cdot & \cdot & \cdot & \cdot & \cdot & \cdot & -1 \\ \cdot & \cdot & -1 & \cdot & \cdot & \cdot & \cdot & \cdot \\ \cdot & \cdot & \cdot & \cdot & \cdot & 1 & \cdot & \cdot \end{bmatrix}. \tag{2.18}$$

These matrices highlight a peculiar trait of the octonions. They satisfy

$$\boxed{e_{R1} = (-e_{L1} + e_{L26} + e_{L34} + e_{L57})/2.} \tag{2.19}$$

This is because for all x in **O**

$$xe_1 = [-e_1x + e_2(e_6x) + e_3(e_4x) + e_5(e_7x)]/2.$$

Likewise,

$$(xe_2)e_6 = [e_1x - e_2(e_6x) + e_3(e_4x) + e_5(e_7x)]/2,$$

so

$$e_{R26} = (e_{L1} - e_{L26} + e_{L34} + e_{L57})/2. \tag{2.20}$$

In general, any product from the right can be reproduced as the sum of products from the left, and visa versa (see chapter 7).

[Note: Because of the cyclic nature of the octonion product, together with the index doubling automorphism, the rules 2.19 and 2.20 are quite general. For example, via doubling 2.20 implies

$$e_{R45} = (e_{L2} - e_{L45} + e_{L61} + e_{L37})/2.$$

Cycling the indices by -1, we get in turn,

$$e_{R34} = (e_{L1} + e_{L26} - e_{L34} + e_{L57})/2$$

(the terms have been arranged in the order they appear in 2.20).]

This property provides a great contrast with the quaternion algebra. If x and b are in $\mathbf{Q}$, then there is no general method of expressing the right product xb of b on x as the sum of left products. The left and right adjoint algebras of $\mathbf{Q}$ on $\mathbf{Q}$ are distinct and commute with one another (assured by the associativity of $\mathbf{Q}$). Also assured by the associativity of $\mathbf{Q}$, the left and right adjoint algebras are each isomorphic to $\mathbf{Q}$ itself.

In the octonion case this is clearly not true. The left and right adjoint algebras ($\mathbf{O}_L$ and $\mathbf{O}_R$) are the same algebra, and this algebra is larger than $\mathbf{O}$ itself. In fact, it is isomorphic to $\mathbf{R}(8)$. It is not difficult to prove that

$$e_{La...bc...d} = -e_{La...cb...d} \tag{2.21}$$

if $b \neq c$, all indices from 1 to 7. So, for example,

$$e_1(e_2(e_3x)) = -e_1(e_3(e_2x)) = e_3(e_1(e_2x)).$$

In addition,

$$e_{Lab...pp...c} = -e_{Lab...c} \tag{2.22}$$

(cancellation of like indices). Together with

$$e_{L7654321} = 1_L \tag{2.23}$$

(the left adjoint identity, also denoted e_{L0}), 2.22 and 2.23 imply that a complete basis for the left=right adjoint algebra of $\mathbf{O}$ consists of elements of the form

$$\boxed{1_L, e_{La}, e_{Lab}, e_{Labc}.} \tag{2.24}$$

This yields 1+7+21+35=64 as the dimension of the adjoint algebra of **O**, also the dimension of **R**(8).

$$\boxed{\mathbf{O}_L = \mathbf{O}_R \simeq \mathbf{R}(8).} \tag{2.25}$$

The 8-dimensional **O** itself is the object, or spinor, space of the adjoint algebra.

Since the left(right) adjoint algebra of **O** is isomorphic to **R**(8), and **O** is 8-dimensional, we call **O** left(right) complete. Again there is a contrast to the other hypercomplex algebras. **Q** is 4-dimensional, but $\mathbf{Q}_L = \mathbf{Q}_R = \mathbf{Q}$ ($\mathbf{Q}_L$ and $\mathbf{Q}_R$ are isomorphic, but distinct). Therefore, since neither is isomorphic to **R**(4), **Q** is neither left nor right complete. However, $\mathbf{Q}_A$, the combined left-right adjoint algebra, with a 16 dimensional basis consisting of the actions

$$x \to x,\ q_i x,\ x q_j,\ q_i x q_j,\ i,j \in \{1,2,3\}, \tag{2.26}$$

is isomorphic to **R**(4). Hence we may call **Q** two-sided complete.

$$\boxed{\begin{array}{l} \mathbf{Q}_L \simeq \mathbf{Q}_R \simeq \mathbf{Q}, \quad \mathbf{Q}_A \simeq \mathbf{R}(4), \\ \mathbf{Q}_L \neq \mathbf{Q}_R. \end{array}} \tag{2.27}$$

(I shall occasionally let the context distinguish q_{Lj} from q_{Rj}, and denote both q_j.)

The case for **C** is even worse. In being both commutative and associative **C** is complete in neither of the above senses.

$$\boxed{\mathbf{C}_L = \mathbf{C}_R = \mathbf{C}_A = \mathbf{C},} \tag{2.28}$$

and all are identical. To generate a full **R**(2) algebra of actions for the 2-dimensional (over the reals) object algebra **C**, one has to resort to actions arising outside the algebra. In particular, the four actions

$$x \to x,\ x^*,\ ix^*,\ ix \tag{2.29}$$

form a basis for a complete **R**(2) algebra of actions on **C** (x^* = the complex conjugate of x: note that the last three actions in 2.30 anticommute).

All of this brings up an interesting point. Because **C** is a subalgebra of **Q** is a subalgebra of **O**, quite often one finds in applications of the division algebras to physics that they are applied singly. Why, after-all, in applying **O** should one separately include **Q** and/or **C**? But the properties that define an algebra or subalgebra are anthropogenic. It is not recorded that there were any higher deities hanging about at the time who could speak with absolute authority on the relevance of these properties to the Truth underlying the design of reality. These adjoint completeness properties set the algebras apart. In the next section, in which the adjoint algebras are identified as Clifford algebras, and the division algebras upon which they act as spinor spaces, these distinctions become more pronounced. And in section 2.6 it will be seen that the fact that "**C** is a subalgebra of **Q** is a subalgebra of **O**" is not as fundamental as it might seem, and in one light it can even be viewed as accidental.

2.3. Clifford Algebras, Spinors.

The left adjoint algebras are

$$\mathbf{C}_L \simeq \mathbf{C},\ \mathbf{Q}_L \simeq \mathbf{Q},\ \mathbf{O}_L \simeq \mathbf{R}(8),$$

which imply the Clifford algebra isomorphisms

$$\mathbf{C}_L \simeq \mathbf{R}_{0,1},\ \mathbf{Q}_L \simeq \mathbf{R}_{0,2},\ \mathbf{O}_L \simeq \mathbf{R}_{0,6}$$

($\mathbf{R}_{p,q}$ the Clifford algebra associated with $(p(+),q(-))$-space-time). In particular, multivector bases for $\mathbf{O}_L \simeq \mathbf{R}_{0,6}$ may be represented as below:

$$\begin{array}{ll} \multicolumn{2}{c}{\underline{\textit{Clifford Algebra } \mathbf{O}_L \simeq \mathbf{R}_{0,6}}} \\ \text{scalar} & 1_L \\ \text{vector} & e_{Lp} \\ \text{2-vector} & e_{Lpq} \\ \text{3-vector} & e_{Lpqr} \\ \text{4-vector} & e_{Lpq7} \\ \text{5-vector} & e_{Lp7} \\ \text{6-vector} & e_{L7}, \end{array} \tag{2.30}$$

$p,q,r \in \{1,2,3,4,5,6\}$ (note that each of these multivector bases is invariant under index doubling).

The spinor space of $\mathbf{R}_{0,6} \simeq \mathbf{R}(8)$ is the space of real 8-spinors, but if represented as in 2.30, the spinor space is just **O** itself, the 8-dimensional object space of $\mathbf{O}_L$. The basis $\{1, e_a, a = 1, ..., 7\}$ of $\mathbf{O}$ is also an orthonormal basis for the spinor space, orthonormal with respect to the inner product

$$< x, y >= (x^\dagger y + y^\dagger x)/2, \tag{2.31}$$

where $x^\dagger$ is the conventional Hermitian transpose antiautomorphism of $\mathbf{O}$, changing the signs of the $e_a, a = 1, ..., 7$. In general, on $\mathbf{C}$, $\mathbf{Q}$, and $\mathbf{O}$,

$$\begin{aligned}(x^0 + \textstyle\sum_1^7 x^a e_a)^\dagger &= x^0 - \textstyle\sum_1^7 x^a e_a,\\ (y^0 + \textstyle\sum_1^3 y^j q_j)^\dagger &= y^0 - \textstyle\sum_1^3 y^j q_j,\\ (z^0 + z^1 i)^\dagger = (z^0 + z^1 i)^* &= z^0 - z^1 i,\end{aligned} \tag{2.32}$$

and 2.31 defines the inner product on each.

Product reversing Hermitian transposition can be extended to $\mathbf{O}_L$ in the obvious way:

$$\begin{aligned} e_{La}^\dagger &= -e_{La},\\ e_{Lab}^\dagger &= e_{Lba} = -e_{Lab},\\ e_{Labc}^\dagger &= -e_{Lcba} = e_{Labc},\end{aligned} \tag{2.33}$$

(a,b,c distinct, from 1 to 7). These actions are consistent with the matrix representations 2.16 and 2.18. Each of those matrices is antisymmetric. It is worth pointing out that even though, for example, $e_{L235}[1] = e_1$, we have $e_{L235} \neq e_{L1}$, and the matrix associated with e_{L235} (in the representation of 2.16 and 2.18) is

$$e_{L235} \rightarrow \begin{bmatrix} . & 1 & . & . & . & . & . & . \\ 1 & . & . & . & . & . & . & . \\ . & . & . & . & . & . & -1 & . \\ . & . & . & . & -1 & . & . & . \\ . & . & . & -1 & . & . & . & . \\ . & . & . & . & . & . & . & -1 \\ . & . & -1 & . & . & . & . & . \\ . & . & . & . & . & -1 & . & . \end{bmatrix},$$

which is symmetric, consistent with $e_{L235}^\dagger = e_{L235}$. With the definition 2.33 in hand one can easily prove that for all x and y in $\mathbf{O}$, and A in $\mathbf{O}_L$,

$$< x, A[y] >=< A^\dagger[x], y > . \tag{2.34}$$

Certain subsets of $\mathbf{O}_L$ close under commutation and so become Lie algebras. The 2-vector subspace of the Clifford algebra 2.30 is an example (see also section 5):

$$\boxed{\{e_{Lpq} : p, q \neq 7\} \to su(4) \simeq so(6) \simeq spin(6)} \tag{2.35}$$

(see section 2.5 and chapters 7 and 8 for a more complete development).

The case for the quaternions, $\mathbf{Q}$, is in some respects more complicated, because $\mathbf{Q}$ is less algebraically complete than $\mathbf{O}$. Multivector bases of $\mathbf{Q}_L$, identified with $\mathbf{R}_{0,2}$, are (r=1,2):

$$\underline{Clifford\ Algebra\ \mathbf{Q}_L \simeq \mathbf{R}_{0,2}}$$

$$\begin{array}{cc} \text{scalar} & 1 \\ \text{vector} & q_{L1},\ q_{L2}, \\ \text{2-vector} & q_{L3}. \end{array} \tag{2.36}$$

The spinor space of $\mathbf{Q}_L$, namely $\mathbf{Q}$ itself, seems larger than it need be, and indeed the elements $\{q_{Ri}, i = 1,2,3\}$ commute with the Clifford algebra 2.36, and so can be considered 'internal' actions relative to the 'external' or geometric actions 2.36. This extra set of internal actions generates an internal $SU(2)$, which will play an important role in what follows. That is, the span of the set $\{q_{Ri}, i = 1,2,3\}$ closes under commutation, and

$$\boxed{\{q_{Ri}, i = 1,2,3\} \to su(2).} \tag{2.37}$$

The inner product of $\mathbf{Q}$ is defined as it was for $\mathbf{O}$. If A is an element of $\mathbf{Q}_L$, then 2.34 is valid in this context as well. In addition, if $x \to xU$ is an internal $SU(2)$ action (from $\mathbf{Q}_R$), then

$$< xU, yU > = < xUU^{\dagger}, y > = < xUU^{-1}, y > = < x, y > . \tag{2.38}$$

As pointed out in 2.27, only the 1-sided adjoint algebras of $\mathbf{Q}$ are algebraically incomplete. $\mathbf{Q}_A \simeq \mathbf{R}(4)$, which is the clifford algebra $\mathbf{R}_{3,1}$. Its spinor space, which we expect to be 4-dimensional, is also $\mathbf{Q}$, which is 4-dimensional. The multivectors of $\mathbf{Q}_A$, identified with $\mathbf{R}_{3,1}$, are representable as follows ($i \in \{1,2,3\}$):

Clifford Algebra $\mathbf{Q}_A \simeq \mathbf{R}_{3,1}$	
scalar	1
vector	$q_{L1}q_{Ri},\ q_{L2}$
2-vector	$q_{Ri},\ q_{L3}q_{Ri}$
3-vector	$q_{L2}q_{Ri},\ q_{L1}$
4-vector	$q_{L3}.$

(2.39)

With respect to $\mathbf{Q}_A$, the 4-dimensional object (spinor) space $\mathbf{Q}$ has no internal degrees of freedom.

2.4. Resolving the Identity of $\mathbf{O}_L$.

An algebraic idempotent, A, is by definition a nonzero element satisfying:

$$A^2 = A. \tag{2.40}$$

A is nontrivial if $A \neq 1$, the identity. In this case

$$A(1-A) = A - A^2 = A - A = 0 \tag{2.41}$$

and

$$(1-A)^2 = 1 - 2A + A^2 = 1 - 2A + A = 1 - A. \tag{2.42}$$

So by 2.42 $1 - A$ is also an idempotent, and by 2.41, A and $1 - A$ are orthogonal.

Because of 2.41, nontrivial idempotents are divisors of zero, hence the identity is the sole idempotent of any division algebra. This applies to $\mathbf{C}$, $\mathbf{Q}$, their left and right adjoint algebras, and to $\mathbf{O}$, but not to $\mathbf{O}_L = \mathbf{O}_R \simeq \mathbf{R}(8)$, which is not a division algebra. In particular, if $a, b, c \in \{1, ..., 7\}$ are distinct, then

$$A = \frac{1}{2}(1_L + e_{Labc}) \tag{2.43}$$

is an idempotent (note: $e_{Labc}^2 = e_{Labcabc} = 1_L$, by 2.21 and 2.22). So is

$$1_L - A = \frac{1}{2}(1_L - e_{Labc}).$$

Certain elements of $\mathbf{O}_L$ are diagonal in the adjoint representation. A basis for these consists of the identity, 1_L, together with the e_{Labc} satisfying

$$e_{Labc}(1) = e_a(e_b e_c) = 1 \tag{2.44}$$

(ie., $q_1 \to e_a$, $q_2 \to e_c$, $q_3 \to e_b$ defines an injection of **Q** into **O**, in which case,

$$e_a(e_b e_c) = (e_a e_b)e_c = e_a e_b e_c = q_1 q_3 q_2 = 1).$$

In particular, define

$$\boxed{I_a = e_{L(3+a)(6+a)(5+a)}} \tag{2.45}$$

(indices from 1 to 7, modulo 7), and let I_0 be the identity. Their adjoint representations are

$$\begin{aligned}
I_0 &= 1_L \to \text{diag}(++++++++),\\
I_1 &= e_{L476} \to \text{diag}(+---+-++),\\
I_2 &= e_{L517} \to \text{diag}(++---+-+),\\
I_3 &= e_{L621} \to \text{diag}(+++---+-),\\
I_4 &= e_{L732} \to \text{diag}(+-++---+),\\
I_5 &= e_{L143} \to \text{diag}(++-++---),\\
I_6 &= e_{L254} \to \text{diag}(+-+-++--),\\
I_7 &= e_{L365} \to \text{diag}(+--+-++-).
\end{aligned} \tag{2.46}$$

(Thus, for example,

$$e_{L517}[e_a] = +e_a, a = 0, 5, 1, 7,$$

$$e_{L517}[e_a] = -e_a, a = 2, 3, 4, 6.)$$

Being diagonal, the I_a clearly commute. They also satisfy

$$I_a I_{a+1} = I_{a+3}, \tag{2.47}$$

$a \in \{1, ..., 7\}$ (had $e_a e_{a+1} = e_{a+3}$ been chosen as the multiplication for **O**, then 2.47 would change to $I_a I_{a+1} = I_{a+5}$, so these choices are in this manner dual to each other (see section 2.6)). Likewise, if $I_a I_b = I_c$, then

$$I_{(2a)} I_{(2b)} = I_{(2c)}. \tag{2.48}$$

Notice that $(2 \cdot 7) = 14 = 7 \bmod 7$, and

$$I_{(2\cdot 7)} = e_{L(2\cdot 3)(2\cdot 6)(2\cdot 5)} = e_{L653} = e_{L365} = I_7$$

(only I_7 is invariant with respect to this index doubling operation).

These elements satisfy some other fascinating properties. For example, for all x in $\mathbf{O}$,

$$I_1[x] = e_{L476}[x] = e_4(e_7(e_6x)) = ((xe_4)e_7)e_6 = e_{R476}[x]. \tag{2.49}$$

(This result is not obvious, but it will be more easily provable after the tools of chapter 7 are developed, which will happen, not coincidentally, in chapter 7.) Moreover,

$$x \to I_a[x] \tag{2.50}$$

is an octonion automorphism (see the next section), so

$$I_a[xy] = I_a[x]I_a[y] \tag{2.51}$$

for all x, y in $\mathbf{O}$.

Finally, the identity of $\mathbf{O}_L$ can be elegantly resolved into orthogonal primitive idempotents using the I_a. A primitive idempotent can not be expressed as the sum of two other idempotents. So, for example,

$$\frac{1}{4}(1 + I_1 + I_2 + I_4) \text{ and } \frac{1}{4}(1 + I_1 - I_2 - I_4)$$

are both idempotents, and so is their sum, $\frac{1}{2}(1 + I_1)$ (note: $I_1I_2 = I_4$). The sum is therefore not a primitive idempotent. As it turns out, neither are the initial two.

Let $i_{ac} = \pm 1$ be the cth sign on the diagonal of the adjoint representation of the I_a in 2.46, c=0,1,...,7. Define

$$\boxed{P_c = \tfrac{1}{8}(\textstyle\sum_{a=0}^{7} i_{ac}I_a).} \tag{2.52}$$

So, for example,

$$P_1 = \frac{1}{8}(I_0 - I_1 + I_2 + I_3 - I_4 + I_5 - I_6 - I_7).$$

These satisfy

$$P_aP_b = \delta_{ab}P_b \tag{2.53}$$

(no sum; $a, b \in \{0, 1, ..., 7\}$), and

$$\sum_{a=0}^{7} P_a = I_0. \tag{2.54}$$

And they are primitive idempotents (in the adjoint representation P_a is the 8x8 real diagonal matrix with 1 in the (a,a) position, all other components zero). Hence they resolve the identity of $\mathbf{O}_L$. Finally, if $x = x^0 + \sum_{a=1}^{7} x^a e_a$, then

$$\begin{aligned} P_0[x] &= x^0, \\ P_a[x] &= x^a e_a \text{ (no sum)},\ a \in \{1,...,7\}. \end{aligned} \tag{2.55}$$

So, for example,

$$(P_0 + P_1 + P_2 + P_6)[x] = x^0 + x^1 e_1 + x^2 e_2 + x^6 e_6.$$

That is, $(P_0+P_1+P_2+P_6)$ is an idempotent projecting from $\mathbf{O}$ a subalgebra isomorphic to $\mathbf{Q}$:

$$(P_0 + P_1 + P_2 + P_6)[\mathbf{O}] \simeq \mathbf{Q}.$$

Likewise, for example,

$$(P_0 + P_1)[\mathbf{O}] \simeq \mathbf{C},$$

and of course,

$$P_0[\mathbf{O}] \simeq \mathbf{R}.$$

2.5. Lie Algebras, Lie Groups, from $\mathbf{O}_L$.

The automorphism group of $\mathbf{O}$ is the 14-dimensional exceptional Lie group G_2. A basis for the Lie algebra of G_2 (denoted, LG_2) is simply represented in $\mathbf{O}_L$:

$$\boxed{\{e_{Lab} - e_{Lcd} : e_a e_b = e_c e_d\} \rightarrow LG_2} \tag{2.56}$$

(this result will be proven in chapter 7).

For example, $e_1 e_5 = e_2 e_3 (= e_7)$, therefore $e_{L15} - e_{L23}$ is an element of LG_2. Therefore, for all real θ,

$$\begin{aligned} G(\theta) &= exp[\tfrac{1}{2}\theta(e_{L15} - e_{L23})] \\ &= \tfrac{1}{2}(1 + I_1) + \tfrac{1}{2}(1 - I_1)cos\theta + \tfrac{1}{2}(e_{L15} - e_{L23})sin\theta \end{aligned} \tag{2.57}$$

is an element of G_2. Let $\theta = \pi$:

$$G(\pi) = \frac{1}{2}(1 + I_1) - \frac{1}{2}(1 - I_1) = I_1$$

(see 2.46), thus confirming that I_1 (hence all the I_a) is an (sign changing) automorphism.

In $\mathbf{O}_R$ the basis is much the same:

$$\{e_{Rab} - e_{Rcd} : e_a e_b = e_c e_d\} \to LG_2,$$

and in fact it is not difficult to prove that

$$e_{Lab} - e_{Lcd} = -(e_{Rab} - e_{Rcd})$$

when $e_a e_b = e_c e_d$.

It is easily verified that $G(\theta)$ defined in 2.57 leaves $1, e_4, e_6$, and e_7 invariant. In general, $e_{Lab} - e_{Lcd} \in LG_2$ will generate G_2 actions that leave 1 and e_f invariant for $f \neq a, b, c, d$.

The stability group of any fixed octonion direction is the 8-dimensional $SU(3) \subset G_2$. Therefore, for example, a basis for the Lie algebra of the stability group of e_7 is:

$$\{e_{Lpq} - e_{Lrs} \in G_2 : p, q, r, s \neq 7\} \to su(3). \tag{2.58}$$

This $su(3)$ is the intersection of LG_2 with $spin(6)$, defined in 2.35, and it will appear as color-$su(3)$ in coming chapters.

Note that the index doubling isomorphism of $\mathbf{O}$ is an element of the $SU(3)$ stability group of e_7, since $e_{2\times 7} = e_{14} = e_7$. This can be explicitly represented as

$$\exp[\tfrac{\pi}{2}(e_{L24} - e_{L63})]\exp[\tfrac{\pi}{2}(e_{L12} - e_{L35})]$$

$$= [\tfrac{1}{2}(1 + I_2) + \tfrac{1}{2}(e_{L24} - e_{L63})][\tfrac{1}{2}(1 + I_1) + \tfrac{1}{2}(e_{L12} - e_{L35})],$$

and index quadrupalling as

$$\exp[\tfrac{\pi}{2}(e_{L35} - e_{L12})]\exp[\tfrac{\pi}{2}(e_{L56} - e_{L41})]$$

$$= [\tfrac{1}{2}(1 + I_1) + \tfrac{1}{2}(e_{L35} - e_{L12})][\tfrac{1}{2}(1 + I_4) + \tfrac{1}{2}(e_{L56} - e_{L41})].$$

Bases for some other subsets of $\mathbf{O}_L$ that close under commutation, and their representation types, are listed below:

$$\{e_{Lab}\} \to so(7), \text{ spinor},$$

$$\{e_{La}, e_{Lbc}\} \text{ and } \{-e_{La}, e_{Lbc}\} \to so(8), \text{ spinor}, \tag{2.59}$$

$$\{e_{La} + e_{Ra},\ e_{Lbc} - e_{Ra} : e_b e_c = e_a\} \to so(8), \text{ vector}.$$

These three examples become more obvious when we tensor $\mathbf{O}_L$ with $\mathbf{R}(2)$. A basis for $\mathbf{R}(2)$ consists of the four matrices

$$\epsilon = \begin{bmatrix} 1 & 0 \\ 0 & 1 \end{bmatrix}, \alpha = \begin{bmatrix} 1 & 0 \\ 0 & -1 \end{bmatrix},$$

$$\beta = \begin{bmatrix} 0 & 1 \\ 1 & 0 \end{bmatrix}, \omega = \begin{bmatrix} 0 & 1 \\ -1 & 0 \end{bmatrix}. \tag{2.60}$$

$\mathbf{O}_L(2) \simeq \mathbf{R}(8) \otimes \mathbf{R}(2) \simeq \mathbf{R}(16)$, which is the Clifford algebra $\mathbf{R}_{8,0}$. Bases for the vector and 2-vector parts of $\mathbf{R}_{8,0}$ are

$$\begin{array}{ll} \text{vector} & \beta, e_{La}\omega, \\ \text{2-vector} & e_{La}\alpha, e_{Lab}\epsilon. \end{array} \tag{2.61}$$

The subspace of 2-vectors closes under commutation and is isomorphic to $so(8)$.

The commutator of a 2-vector with a vector is a vector, this giving rise to the vector representation of $so(8)$. Let $x_L = x^0 + \sum_1^7 x^a e_{La}$, and if A is in $\mathbf{O}_L$, let $A^\dagger$ be the conjugation defined in 2.33. Then

$$\left[\begin{bmatrix} e_{La} & 0 \\ 0 & -e_{La} \end{bmatrix}, \begin{bmatrix} 0 & x_L \\ x_L^\dagger & 0 \end{bmatrix}\right] = \begin{bmatrix} 0 & (e_{La}x_L + x_L e_{La}) \\ (same)^\dagger & 0 \end{bmatrix}.$$

The operation of $e_{La}x_L + x_L e_{La}$ on the identity of $\mathbf{O}$ yields (let $x = (x^0 + \sum_1^7 x^a e_a) \in \mathbf{O}$)

$$\begin{aligned} e_{La}x_L[1] + x_L e_{La}[1] &= e_a(x1) + x(e_a 1) \\ &= e_a x + x e_a = (e_{La} + e_{Ra})[x]. \end{aligned} \tag{2.62}$$

Likewise

$$\left[\begin{bmatrix} e_{Lbc} & 0 \\ 0 & e_{Lbc} \end{bmatrix}, \begin{bmatrix} 0 & x_L \\ x_L^\dagger & 0 \end{bmatrix}\right] = \begin{bmatrix} 0 & (e_{Lbc}x_L + x_L e_{Lbc}) \\ (same)^\dagger & 0 \end{bmatrix},$$

and

$$\begin{aligned} (e_{Lbc}x_L + x_L e_{Lbc})[1] &= e_b(e_c(x1)) - x(e_b(e_c 1)) \\ &= e_{Lbc}[x] - e_{Ra}[x], \end{aligned} \tag{2.63}$$

where $e_a = e_b(e_c 1) = e_b e_c$. This justifies the claim that the third of the $so(8)$ representations in 2.59 is vector.

The other two are simpler. The spinor space of $\mathbf{O}_L(2)$ is $\mathbf{O}^2$, 2x1 matrices over $\mathbf{O}$. Let X and Y be in $\mathbf{O}$. The 2-vector ($so(8)$) actions from 2.61 on $\mathbf{O}^2$ take the forms

$$\begin{aligned} \begin{bmatrix} e_{La} & 0 \\ 0 & -e_{La} \end{bmatrix} \begin{bmatrix} x \\ y \end{bmatrix} &= \begin{bmatrix} e_{La}[x] \\ -e_{La}[y] \end{bmatrix}, \\ \begin{bmatrix} e_{Lbc} & 0 \\ 0 & e_{Lbc} \end{bmatrix} \begin{bmatrix} x \\ y \end{bmatrix} &= \begin{bmatrix} e_{Lbc}[x] \\ e_{Lbc}[y] \end{bmatrix}, \end{aligned} \tag{2.64}$$

thereby justifying the two spinor representations in 2.59. Thus the two spinor representations and the vector representation all act on 8-dimensional $\mathbf{O}$. These three $so(8)$ representations are also linked by what is known as $so(8)$ triality. This concept will be developed more completely in section 7.3.

2.6. From Galois Fields to Division Algebras: An Insight.

The real numbers are the paradigm for mathematical field theory. There is addition (and subtraction), an additive identity, 0, and every element x has an additive inverse, $-x$. There is multiplication (and division), a multiplicative identity, 1, and every element $x \neq 0$ has a multiplicative inverse, x^{-1}. Multiplication by zero gives zero, and for all $x \neq 0$ and $y \neq 0$, we also have $xy \neq 0$ (no divisors of zero). Finally, $xy = yx$ (commutative), and $x(yz) = (xy)z$ (associative).

$\mathbf{R}$ is an infinite field, but there also exist finite fields. For any prime p there exist (unique up to isomorphism) fields of order p^k for all $k = 1, 2, 3, \dots$, denoted $GF(p^k)$ (G for Galois, their ill-fated founder, F for field). For no other positive integers are there fields of that order.

The p^k elements of $GF(p^k)$ are easily written: $\{0, 1, h, h^2, \dots, h^{p^k-2}\}$, where h is the multiplicative generator. That is, the multiplication of $GF(p^k)$ is cyclic and for all $x \neq 0$ in $GF(p^k)$,

$$x^{p^k-1} - 1 = 0 \tag{2.65}$$

(ie., $h^{p^k-1} = 1$).

All that remains then is to construct an addition table for $GF(p^k)$ consistent with its being a field. This problem can be reduced to finding what is called a Galois sequence for $GF(p^k)$, which consists of $p^k - 1$ elements of Z_p

(the integers modulo p). Its further properties can be best illustrated by an example. (Mathematicians have a more elaborate development in terms of polynomials and quotient modules; the elements of a Galois sequence appear in that context as coefficients of a polynomial.)

$[\ 0\ \ 1\ \ 1\ \ 2\ \ 0\ \ 2\ \ 2\ \ 1\ ]$ is a Galois sequence for $GF(3^2 = 9)$. We identify it with $h^0 = 1$, the multiplicative identity of $GF(9)$, and we'll identify its kth cyclic permutation with h^k. That is,

$$h^1 = [\ 1\ \ 0\ \ 1\ \ 1\ \ 2\ \ 0\ \ 2\ \ 2\ ],$$

$$h^2 = [\ 2\ \ 1\ \ 0\ \ 1\ \ 1\ \ 2\ \ 0\ \ 2\ ],$$

$$h^3 = [\ 2\ \ 2\ \ 1\ \ 0\ \ 1\ \ 1\ \ 2\ \ 0\ ],$$

$$h^4 = [\ 0\ \ 2\ \ 2\ \ 1\ \ 0\ \ 1\ \ 1\ \ 2\ ],$$

$$h^5 = [\ 2\ \ 0\ \ 2\ \ 2\ \ 1\ \ 0\ \ 1\ \ 1\ ],$$

$$h^6 = [\ 1\ \ 2\ \ 0\ \ 2\ \ 2\ \ 1\ \ 0\ \ 1\ ],$$

$$h^7 = [\ 1\ \ 1\ \ 2\ \ 0\ \ 2\ \ 2\ \ 1\ \ 0\ ],$$

$$h^8 = [\ 0\ \ 1\ \ 1\ \ 2\ \ 0\ \ 2\ \ 2\ \ 1\ ], \tag{2.66}$$

where $h^8 = h^0 = 1$ gets us back to where we started from (any cyclic permutation of the initial sequence would have been a valid starting point). Notice that the first $k = 2$ elements of each sequence are unique, and can be used as labels for the elements (we are using instead the exponents). And notice that by adjoining to this collection the zero sequence, $0 = [\ 0\ \ 0\ \ 0\ \ 0\ \ 0\ \ 0\ \ 0\ \ 0\ ]$, we have a set of $p^k = 3^2 = 9$ vectors (sequences), each $p^k - 1 = 3^2 - 1 = 8$-dimensional over $Z_p = Z_3$, and that the set is closed with respect to Z_3 vector addition. Let $+_p$ represent addition modulo p. Then for example,

$$h^2 +_3 h^4 = [\ 2\ \ 1\ \ 0\ \ 1\ \ 1\ \ 2\ \ 0\ \ 2\ ] +_3 [\ 0\ \ 2\ \ 2\ \ 1\ \ 0\ \ 1\ \ 1\ \ 2\ ]$$

$$= [\ 2\ \ 0\ \ 2\ \ 2\ \ 1\ \ 0\ \ 1\ \ 1\ ] = h^5.$$

A full addition table for $GF(9)$ resulting from this sequence is listed below:

0	h^1	h^2	h^3	h^4	h^5	h^6	h^7	h^8
h^1	h^5	h^8	h^4	h^6	0	h^3	h^2	h^7
h^2	h^8	h^6	h^1	h^5	h^7	0	h^4	h^3
h^3	h^4	h^1	h^7	h^2	h^6	h^8	0	h^5
h^4	h^6	h^5	h^2	h^8	h^3	h^7	h^1	0
h^5	0	h^7	h^6	h^3	h^1	h^4	h^8	h^2
h^6	h^3	0	h^8	h^7	h^4	h^2	h^5	h^1
h^7	h^2	h^4	0	h^1	h^8	h^3	h^3	h^6
h^8	h^7	h^3	h^5	0	h^2	h^1	h^4	h^4

(2.67)

(recall that $h^8 = 1$). Note that

$$h^k +_3 h^k = h^{k+4}$$

and

$$h^k +_3 h^k +_3 h^k = h^k +_3 h^{k+4} = 0.$$

Also,

$$h^k +_3 h^{k+1} = h^{k+7}.$$

Because for any x and y in any $GF(3^m)$,

$$(x +_3 y)^3 = x^3 +_3 y^3, \tag{2.68}$$

cubing the last equation above results in

$$h^k +_3 h^{k+3} = h^{k+5}$$

(exponents are taken modulo 8 from 1 to 8, and although strictly speaking the exponents k cube to $3k$, because 3 and 8 are relatively prime we are allowed to replace $3k$ by k in constructing new addition rules), and cubing this leads back to $h^k +_3 h^{k+1} = h^{k+7}$. There is also,

$$h^k +_3 h^{k+2} = h^{k+3},$$

which cubed yields,

$$h^k +_3 h^{k+6} = h^{k+1},$$

and finally

$$h^k +_3 h^{k+5} = h^{k+2},$$

which cubed yields,

$$h^k +_3 h^{k+7} = h^{k+6}.$$

Of more interest to us here are the fields $GF(2^n), n = 1, 2, 3$. In particular, a Galois sequence

$$\text{for } GF(2^1) \text{ is } [\ 1\ ],$$

$$\text{for } GF(2^2) \text{ is } [\ 0\ \ 1\ \ 1\ ],$$

$$\text{for } GF(2^3) \text{ is } [\ 0\ \ 0\ \ 1\ \ 0\ \ 1\ \ 1\ \ 1\ ].$$

In this last case we define

$$\begin{aligned}
e^1 &= [\ 1\ \ 0\ \ 0\ \ 1\ \ 0\ \ 1\ \ 1\ ],\\
e^2 &= [\ 1\ \ 1\ \ 0\ \ 0\ \ 1\ \ 0\ \ 1\ ],\\
e^3 &= [\ 1\ \ 1\ \ 1\ \ 0\ \ 0\ \ 1\ \ 0\ ],\\
e^4 &= [\ 0\ \ 1\ \ 1\ \ 1\ \ 0\ \ 0\ \ 1\ ],\\
e^5 &= [\ 1\ \ 0\ \ 1\ \ 1\ \ 1\ \ 0\ \ 0\ ],\\
e^6 &= [\ 0\ \ 1\ \ 0\ \ 1\ \ 1\ \ 1\ \ 0\ ],\\
e^7 &= [\ 0\ \ 0\ \ 1\ \ 0\ \ 1\ \ 1\ \ 1\ ].
\end{aligned} \tag{2.69}$$

Addition in this case can also be completely described by cyclic equations in the e^a. To begin with,

$$e^a +_2 e^a = 0 \tag{2.70}$$

(every element is its own additive inverse, and $+_2$ is the binary vector addition). Also,

$$e^a +_2 e^{a+1} = e^{a+5}. \tag{2.71}$$

Since in this case ($p = 2$)

$$(x +_2 y)^2 = x^2 +_2 y^2, \tag{2.72}$$

squaring the above addition rule leads to a new rule,

$$e^a +_2 e^{a+2} = e^{a+3}, \tag{2.73}$$

and squaring this leads to

$$e^a +_2 e^{a+4} = e^{a+6} \tag{2.74}$$

(exponents are taken modulo 7 from 1 to 7).

The link of $GF(8)$ to the octonions should now be obvious. The matrix of signs in 2.10, used to construct an octonion multiplication, could have been replaced by the following matrix of elements of Z_2 (ie., 0's and 1's):

$$O' = \begin{bmatrix} 0 & 0 & 0 & 0 & 0 & 0 & 0 & 0 \\ 0 & 1 & 0 & 0 & 1 & 0 & 1 & 1 \\ 0 & 1 & 1 & 0 & 0 & 1 & 0 & 1 \\ 0 & 1 & 1 & 1 & 0 & 0 & 1 & 0 \\ 0 & 0 & 1 & 1 & 1 & 0 & 0 & 1 \\ 0 & 1 & 0 & 1 & 1 & 1 & 0 & 0 \\ 0 & 0 & 1 & 0 & 1 & 1 & 1 & 0 \\ 0 & 0 & 0 & 1 & 0 & 1 & 1 & 1 \end{bmatrix}. \tag{2.75}$$

Note that $(-1)^{O'_{ab}} = O_{ab}$ (see 2.10), so if we define

$$O'_a * O'_b = (-1)^{O'_{ab}}[O'_a +_2 O'_b], \tag{2.76}$$

then we have once again created an octonion product, where this time the rows of O' are identified with the basis of the octonions. Note! We have used $GF(8)$ addition to create an octonion multiplication. The first row of O' is the multiplicative identity of $\mathbf{O}$, and we must create a new 0 to play the role of the additive identity of $\mathbf{O}$. With respect to $\mathbf{O}$ addition, the rows of O' are now treated as linearly independent, a basis for a real algebra.

Relabel the rows of O' as $e_a, a = 0, 1, ..., 7$. So the exponents of $GF(8)$ in 2.69 have become the subscripts of the octonions. Because the octonion product (now denoted just $e_a e_b$) is derived directly from the $GF(8)$ addition, the exponent rules 2.71,73,74 are valid for the octonion product, the rules now applied to subscripts (see 2.4,7). In addition, the index doubling automorphism for the octonions (2.6) is now seen to follow from 2.72.

[Note: The sum rules 2.71,73,74 for $GF(8)$ correspond to 2.4,7, but in general we can only make such correspondences up to a sign. For example, while it is true in $GF(8)$ that $e^a +_2 e^{a+5} = e^{a+1}$, in $\mathbf{O}$ we have $e_a e_{a+5} = -e_{a+1}$. Index doubling is also tricky, and in $\mathbf{Q}$ it works out slightly differently.]

[Also note: In $GF(8)$, $e^7 = e^0 = 1$. The reason that it was listed as e^7 in 2.69 is to make the correspondence $e^7 \rightarrow e_7$ of $GF(8)$ to $\mathbf{O}$. Therefore, since $e^0 = e^7$, we have $e^0 \rightarrow e_7$, too! That is, we use e_0 to denote the octonion identity, 1, and it has no correspondence to any power of $e^1 \in GF(8)$.

In [10] the identity is given the subscript ∞, so we could write $e_\infty = 1$. This notation has distinct advantages, but also for our purposes disadvantages. I'd apologize for the confusion of all this were it not that this section is only intended to highlight the naturalness of the chosen **O** product, and this material will not play a part in what follows.]

[Finally note: the transpose of O' also results in a valid $GF(8)$ addition and **O** multiplication. In this case, however, $e_a e_{a+1} = -e_{a+3}$ in **O**. Except for the sign change, this is the dual multiplication mentioned in section 2.4. If we replace 2.76 by

$$O'_a * O'_b = (-1)^{O'_{ba}}[O'_a +_2 O'_b],$$

we generate the **O** multiplication rule, $e_a e_{a+1} = -e_{a+5}$; and if we use the transpose of O', the rule $e_a e_{a+1} = e_{a+3}$. So strictly speaking there are 4 cyclic octonionic multiplication rules, 2 for the quaternions, and 1 for the complexes.]

Having made the correspondence between $GF(8)$ addition and **O** multiplication, one is naturally led to consider the role of $GF(8)$ multiplication in **O**. Since in $GF(8)$, $e^a e^b = e^{a+b}$, this operation on the indices of **O** is just a cyclic shift (of the index a for $a = 1, ..., 7$; e_0 is left unaltered). Let S be the **O** automorphism that shifts the indices of $e_b, b = 1, ..., 7$ by 1. So S^a shifts the **O** indices by a, and $S^7 = S^0$ is the identity map. Let ϕ be the zero map, mapping all $x \in \mathbf{O}$ to 0. Obviously this collection of eight maps can be made into the field $GF(8)$ if given the appropriate addition. This may or may not be of interest, but this is as far down that road as I am willing to go at present. (Note that although S^a is an element of G_2, unlike the index doubling automorphism it is not an element of the stability $SU(3)$ subgroup for any e_b except when $a = 0$.)

In the quaternion case one makes a correspondence with $GF(4)$. Everything works out much the same, save that 2.72 doesn't give rise to as simple a relation in **Q** as it did in **O**. By inspection we see in this case that if $q_i q_j = q_k$, then

$$q_{(2i)} q_{(2j)} = -q_{(2k)} \tag{2.77}$$

(therefore index doubling can be regarded as an antiautomorphism in this case). Index quadrupalling gets us back to $q_i q_j = q_k$, since 4=1 mod 3. Hence in **O**, $e_a e_b = e_c$ could not imply $e_{(2a)} e_{(2b)} = -e_{(2c)}$, since $2^3 = 8 = 1$

mod 7, and three (an odd number of) applications of index doubling must get us back to $e_a e_b = e_c$.

The binary matrix generating both **Q** multiplication and $GF(4)$ addition is

$$Q' = \begin{bmatrix} 0 & 0 & 0 & 0 \\ 0 & 1 & 0 & 1 \\ 0 & 1 & 1 & 0 \\ 0 & 0 & 1 & 1 \end{bmatrix}. \tag{2.78}$$

In both O' (see 2.75) and Q', the first row of each after the zeroth must be either the one shown, or the first row of the respective transposes, for algebras isomorphic to **O** and **Q** to result from the process outlined. In particular, consider

$$B = \begin{bmatrix} 0 & 0 & 0 & 0 \\ 0 & 0 & 1 & 1 \\ 0 & 1 & 0 & 1 \\ 0 & 1 & 1 & 0 \end{bmatrix}. \tag{2.79}$$

$[B_{11}\ B_{12}\ B_{13}] = [0\ 1\ 1]$ is also a Galois sequence for $GF(4)$, but in this case the algebra multiplication

$$B_i * B_j = (-1)^{B_{ij}}[B_i +_2 B_j] \tag{2.80}$$

does not result in **Q**, but rather an algebra isomorphic to that generated by the adjoint elements, $q_{L1}q_{R3}$, $q_{L2}q_{R2}$, $q_{L3}q_{R1}$.

Addition on $GF(2^n)$ can be turned into an algebra multiplication in the way outlined for $n > 3$ as well. For example, let

$$g^1 = [\ 0\ \ 0\ \ 0\ \ 1\ \ 0\ \ 0\ \ 1\ \ 1\ \ 0\ \ 1\ \ 0\ \ 1\ \ 1\ \ 1\ \ 1\ ],$$

15-dimensional over Z_2. This is a Galois sequence for $GF(16)$, and it can be used to construct a new 16-dimensional algebra, extending the sequence, **R**, **C**, **Q**, **O** (this is distinct from the Cayley-Dickson prescription, which is founded on the inclusion property, and in fact **O** is not a subalgebra of this new 16-dimensional algebra, which is noncommutative, nonassociative, and nonalternative).

One final path down which I have no intention of travelling far: we should be able to construct algebras in like manner from any $GF(p^n)$, for any prime p. For example, take the $h^k, k = 1, ..., 8$, in $GF(9)$ listed in 2.66, and map them to $h_k, k = 1, ..., 8$, part of a basis for a new algebra. Map the zero sequence to 1, completing the basis. Form the stacked sequences in 2.66 into a matrix, H (8×8). If $h^i +_3 h^j = h^k$ in $GF(9)$, then define

$$h_i h_j = exp[2\pi i H_{ij}/3] h_k. \tag{2.81}$$

If $j - i = 4 \bmod 8$, then replace h_k by 1. At this point I'm just spewing out ideas without a clear notion of their interest or viability, so I'll shift directions a bit in hopes of bringing order out of chaos.

It would seem in light of the material presented to this point in this section that the division algebras are four out of an infinite collection of possible algebras constructable in like manner. Worse, it *is* a collection, not a sequence. Highlighting this is the fact that the first rows of Q' and O' (ignoring the intitial 0's) had to be [1 0 1] and [1 0 0 1 0 1 1] for **Q** and **O** with the multiplication rules we are adopting to result. Completely different algebras result from most of the other cyclic permutations of these sequences.

We could just as well have begun with the dual sequences [1 1 0] and [1 1 1 0 1 0 0] (ending with the same element, but reverse the order of the sequence). These sequences also give rise to **Q** and **O**, and they are Galois sequences for $GF(4)$ and $GF(8)$. They are in addition *quadratic residue codes* of lengths 3 and 7 over $GF(2)$ (see **[10]**). For example, the quadratic residues modulo 7 are $0^2 = 7^2 = 0$, $1^2 = 6^2 = 1$, $2^2 = 5^2 = 4$, $3^2 = 4^2 = 2$, so confusingly renumbering the positions of the sequence [1 1 1 0 1 0 0] from 0 to 6, we see that the 1's appear in the 0, 1, 2, and 4 positions, which are determined by the quadratic residues. Likewise, modulo 3, $0^2 = 3^2 = 0$, $1^2 = 2^2 = 1$, and the 1's of [1 1 0] appear in the 0 and 1 positions. The quadratic residue code of length 1 over $GF(2)$ is [1], also the Galois sequence of $GF2$, and associated with **C**.

There are *no* other examples of quadratic residue codes over $GF(2)$ that correspond to Galois sequences. To even have a chance we must have a code of length $2^k - 1$, and $2^k - 1$ must be prime. So 15 is out. The quadratic residue code of length 31 is

$$[1110110111100010101110000100100]; \tag{2.82}$$

and a Galois sequence, equal to 2.82 in the first seven positions, is

$$[1110110001111100110100100001010]. \tag{2.83}$$

Let U be the 31x31 matrix formed of the first of these sequences and all its cyclic permutations, and let V be the 31x31 matrix formed from the second. The first has the nice property shared by all quadratic residue codes over $GF(2)$ that

$$(-1)^{U_{ab}} = -(-1)^{U_{ba}}, a \neq b. \tag{2.84}$$

In the $2^2-1=3$ and $2^3-1=7$ cases this gives rise to the noncommutativity among the imaginary basis elements ($\neq 1$) of **Q** and **O**, which together with

$$(-1)^{U_{aa}} = -1 \tag{2.85}$$

ensures that **Q** and **O** are division algebras (replace U by the appropriate 3x3 and 7x7 matrices). Unfortunately the rows of U are not closed under Z_2 addition. Those of V are:

$$V_a +_2 V_b = V_c, \tag{2.86}$$

for some c, with $a \neq b$. Without this property the matrix will not give rise to an algebra in the manner outlined (of course, an initial column and row of 0's has to be tacked on). Requiring of our generating sequences that they be both Galois and quadratic residue is a heavy restriction, and the division algebras are the only algebras that result. They once again form a sequence, but it is definitely finite.

3. Tensor Algebras.

3.1. Tensoring Two: Clifford Algebras and Spinors.

Continuing the policy of viewing adjoint division algebras as Clifford algebras, and the division algebras themselves as spinor spaces, define

$$\boxed{\mathbf{P}_L = \mathbf{C}_L \otimes \mathbf{Q}_L,}$$

which is isomorphic to $\mathbf{C}(2)$ and is identified with the Clifford algebra $\mathbf{R}_{3,0}$. Multivector bases for $\mathbf{P}_L \simeq \mathbf{R}_{3,0}$ are listed below (j=1,2,3):

$$\underline{Clifford\ Algebra\ \mathbf{P}_L \simeq \mathbf{R}_{3,0}}$$

$$\begin{array}{ll} \text{scalar} & 1 \\ \text{vector} & iq_{Lj} \\ \text{2-vector} & q_{Lj} \\ \text{3-vector} & i. \end{array} \tag{3.1}$$

As usual the subspace of 2-vectors closes under commutation. In this case it is isomorphic to

$$\boxed{so(3) \simeq su(2) \simeq spin(3).}$$

Let

$$U = exp(\sum_j \theta^j q_{Lj}), \text{ and } X = \sum_j X^j q_{Lj},$$

and let $\psi \in \mathbf{C} \otimes \mathbf{Q}$, the spinor space of $\mathbf{P}_L$. Then the action

$$X \rightarrow UXU^{-1}$$

is an $SO(3)$ rotation of the Euclidean vector $\{X^j\}$, and the action

$$\psi \rightarrow U\psi$$

is and $SU(2) = Spin(3)$ rotation of the spinor ψ.

The object space of $\mathbf{C}(2)$ is the space of 2x1 complex Pauli spinors. The object (spinor) space of $\mathbf{P}_L$ is $\mathbf{P} = \mathbf{C} \otimes \mathbf{Q}$, which is 4-dimensional over the complexes, hence it has twice the dimension of the space of Pauli spinors. With the help of orthogonal Hermitian idempotents, $\mathbf{P}$ can be decomposed into the sum of two Pauli spinor spaces. Let $\vec{x} = \sum_1^3 x^j q_j$, x^j real, such that $\vec{x}^2 = -1$. Define

$$\boxed{\lambda_\pm = \tfrac{1}{2}(1 \pm i\vec{x}),} \tag{3.2}$$

These satisfy

$$\lambda_r \lambda_s = \delta_{rs} \lambda_r, \tag{3.3}$$

and

$$\lambda_+ + \lambda_- = 1. \tag{3.4}$$

As they are primitive idempotents, by 3.3 and 3.4 they constitute a resolution of the identity of $\mathbf{P}$. The subspaces $\mathbf{P}\lambda_\pm$ are stable under the action of $\mathbf{P}_L$. Each is 2-dimensional over $\mathbf{C}$ and each transforms as a Pauli spinor with respect to $\mathbf{P}_L$.

In particular, if we set

$$\vec{x} = x^2 q_2 + x^3 q_3 \tag{3.5}$$

(which is done for the purposes of chapter 4), then $\{q_1, \vec{x}, q_1\vec{x}\}$ form an orthonormal basis for the hypercomplex part of $\mathbf{Q}$. Since

$$\boxed{\vec{x}\lambda_\pm = \tfrac{1}{2}(\vec{x} \mp i) = \mp i \lambda_\pm,} \tag{3.6}$$

any element z of $\mathbf{P}\lambda_\pm$ will, with this choice for $\vec{x}$, reduce to the form

$$z = (z^0 + z^1 q_1)\lambda_\pm, \tag{3.7}$$

$z^0, z^1 \in \mathbf{C}$. Therefore, for example, the action

$$\begin{aligned} z \to \quad & i q_1 \vec{x} z = i q_1 \vec{x}(z^0 + z^1 q_1)\lambda_\pm \\ &= q_1 (z^0 - z^1 q_1) i \vec{x} \lambda_\pm \\ &= \pm (z^1 + z^0 q_1)\lambda_\pm \end{aligned}$$

is equivalent to the action

$$\begin{bmatrix} z^0 \\ z^1 \end{bmatrix} \to \pm \begin{bmatrix} 0 & 1 \\ 1 & 0 \end{bmatrix} \begin{bmatrix} z^0 \\ z^1 \end{bmatrix} = \pm \begin{bmatrix} z^1 \\ z^0 \end{bmatrix},$$

which is the action of a Pauli matrix on a Pauli spinor. All of the elements of

$$\{1_L,\ i_L,\ q_{L1},\ iq_{L1},\ \vec{x}_L = x^2 q_{L2} + x^3 q_{L3},\ i\vec{x}_L,\ q_{L1}\vec{x}_L,\ iq_{L1}\vec{x}_L\}$$

can be be similarly identified with elements of $\mathbf{C}(2)$.

Note that the subspaces $\{\mathbf{P}\lambda_\pm\}$ transform as an $su(2)$ doublet with respect to the right action of the q_j. For example,

$$Xe^{\theta\vec{x}} = X\lambda_+ e^{\theta\vec{x}} + X\lambda_- e^{\theta\vec{x}} = X\lambda_+ e^{-i\theta} + X\lambda_- e^{i\theta}. \tag{3.8}$$

So the action $X \to X\vec{x} = X\lambda_+(-i) + X\lambda_-(i)$ can be written as the matrix action

$$[\ X\lambda_+ \quad X\lambda_-\ ] \to [\ X\lambda_+ \quad X\lambda_-\ ] \begin{bmatrix} -i & 0 \\ 0 & i \end{bmatrix}, \tag{3.9}$$

which is recognizably $su(2)$.

Note that the action 3.8 is an element of $\mathbf{Q}_R$, and that $\mathbf{Q}_R$ commutes with $\mathbf{P}_L = \mathbf{R}_{3,0}$, which is the "external" geometric (Clifford) algebra. Any action on the spinor space which is not an element of the Clifford algebra, and commutes with the action of the Clifford algebra, is considered "internal" relative to that algebra. The set of elements of $\mathbf{Q}_R$ of unit length is isomorphic to $SU(2)$ (the action 3.8 arises from this group), hence this $SU(2)$ is inherently internal.

3.2. Tensoring Two: Spinor Inner Product.

Let A and B be elements of $\mathbf{P}$. Each decomposes with respect to $\lambda_\pm$ into a pair of Pauli spinors:

$$\begin{aligned} A &= A\lambda_+ + A\lambda_-, \\ B &= B\lambda_+ + B\lambda_-. \end{aligned} \tag{3.10}$$

Consider the expression

$$\begin{aligned} (A\lambda_\pm)^\dagger(B\lambda_\pm) &= \lambda_\pm A^\dagger B \lambda_\pm \\ &=< A^\dagger B >_\pm \lambda_\pm, \end{aligned} \tag{3.11}$$

where the $< A^\dagger B >_\pm$ are complex. This follows from the fact that all $X \in \mathbf{P}$ may be expressed as

$$X = X^0 + X^1 iq_1 + X^2 iq_1\vec{x} + X^3 i\vec{x},$$

where $X^m \in \mathbf{C}$, $m = 0, 1, 2, 3$. Since

$$\lambda_\pm q_1 \lambda_\pm = q_1 \lambda_\mp \lambda_\pm = 0,$$

$$\lambda_\pm q_1 \vec{x} \lambda_\pm = q_1 \vec{x} \lambda_\mp \lambda_\pm = 0,$$

and

$$\lambda_\pm i\vec{x} \lambda_\pm = \pm\lambda_\pm,$$

therefore

$$\lambda_\pm X \lambda_\pm = (X^0 \pm X^3)\lambda_\pm = X_\pm \lambda_\pm.$$

In fact, $< A^\dagger B >_\pm$ is just the conventional complex inner product of the Pauli spinor $A\lambda_\pm$ with the Pauli spinor $B\lambda_\pm$. We define the real inner product of A and B to be

$$< A, B > = \frac{1}{4}\{[< A^\dagger B >_+ + < A^\dagger B >_-] + [same]^\dagger\}. \tag{3.12}$$

$< A, B >$ is just the real part of $A^\dagger B$ (or $B^\dagger A$). To see this, let $X \in \mathbf{P}$, and define the four components of X with respect to the $\lambda_\pm$ by $\lambda_s X \lambda_r$ $(r, s = \pm)$. In particular, the diagonal components are

$$\lambda_\pm X \lambda_\pm = X_\pm \lambda_\pm, \tag{3.13}$$

where the $X_\pm$ are complex (so $< A^\dagger B >_\pm$ are the diagonal components of $A^\dagger B$). Define the trace of X with respect to the $\lambda_\pm$ by

$$tr(X) = X_+ + X_-. \tag{3.14}$$

Let $X \to X^-$ be the antiautomorphism on $\mathbf{P}$ that reverses the signs of $q_i, i = 1, 2, 3$, but not the sign of i (so $X^- = X^{\dagger *}$, Hermitian conjugation followed by complex conjugation). Therefore, $\lambda_{\pm}^{-} = \lambda_{\mp}$. Then

$$\begin{aligned} X_+ + X_- &= X_+(\lambda_+ + \lambda_-) + X_-(\lambda_+ + \lambda_-) \\ &= X_+\lambda_+ + (X_+\lambda_+)^- + (X_-\lambda_-)^- + X_-\lambda_- \\ &= \lambda_+ X\lambda_+ + (\lambda_+ X\lambda_+)^- + (\lambda_- X\lambda_-)^- + \lambda_- X\lambda_- \\ &= \lambda_+ X\lambda_+ + \lambda_- X^-\lambda_- + \lambda_+ X^-\lambda_+ + \lambda_- X\lambda_- \\ &= \lambda_+(X + X^-)\lambda_+ + \lambda_-(X + X^-)\lambda_- \\ &= (X + X^-)\lambda_+ + (X + X^-)\lambda_- \\ &= (X + X^-)(\lambda_+ + \lambda_-) \\ &= X + X^-, \end{aligned} \tag{3.15}$$

where $X + X^-$ is clearly complex, so $\lambda_{\pm}(X + X^-)\lambda_{\pm} = (X + X^-)\lambda_{\pm}$. Therefore

$$tr(X) = X_+ + X_- = X + X^-$$

is just twice the complex part of X.

Define

$$M_0(X) = X,$$

the identity map,

$$M_1(X) = X^* = (X^\dagger)^-,$$

an automorphic involution (note, * is complex conjugation on **P**, not on **C**(2)). The inner product may now be constructively expressed:

$$\boxed{< A, B >= \tfrac{1}{4}\{[\textstyle\sum_{rs} M_r((A\lambda_s)^\dagger(B\lambda_s))] + [same]^\dagger\}} \tag{3.16}$$

$(r = 0, 1, s = \pm)$. This follows from the fact that

$$M_1(\lambda_{\pm}) = \lambda_{\mp},$$

so that, for example,

$$\textstyle\sum_{rs} M_r(\lambda_s(X + X^*)\lambda_s) = \sum_{rs} M_r((X_s + X_s^*)\lambda_s)$$

$$= \textstyle\sum_s[(X_s + X_s^*)\sum_r M_r(\lambda_s)] = \sum_s(X_s + X_s^*) = 4Re(X).$$

Another, simpler, expression for $< A, B >$ is the following:

$$< A, B >= \frac{1}{4}\{[(A^{\dagger}B) + (A^{\dagger}B)^{-}] + [same]^{\dagger}\}.$$

The expression 3.16 has the advantage that it highlights the connection between the Pauli spinor parts of A and B. In section 4 this form will be generalized to the more elaborate case of $\mathbf{C} \otimes \mathbf{Q} \otimes \mathbf{O}$. In both cases it facilitates the ultimate connection to physics, to be made in chapter 4, and the derivation of important symmetries.

LINK TO INTERNAL SYMMETRY

In particular, if in 3.16 we replace $\lambda_{\pm}$ by $\kappa_{\pm}$ satisfying

$$\boxed{\kappa_r^{\dagger}\kappa_s = \delta_{rs}\lambda_r,} \tag{3.17}$$

then the result is the same. This is so because by 3.17 the components of the identity of $\mathbf{P}$ with respect to $\kappa_{\pm}$ are the same as the components with respect to $\lambda_{\pm}$ (see 3.13), namely, δ_{rs} .

The symmetry of 3.17 is

$$\boxed{U(1) \times U(1) \times SU(2).}$$

Note first that

$$\begin{aligned} (\kappa_r\lambda_s)^{\dagger}(\kappa_r\lambda_s) &= \lambda_s(\kappa_r^{\dagger}\kappa_r)\lambda_s \\ &= \lambda_s\lambda_r\lambda_s \qquad (3.18) \\ &= \delta_{rs}\lambda_r. \end{aligned}$$

If $X^{\dagger}X = 0$, $X \in \mathbf{P}$, then $X = 0$. Therefore,

$$\kappa_r = \kappa_r\lambda_+ + \kappa_r\lambda_- = \kappa_r\lambda_r.$$

That is,

$$\boxed{\kappa_{\pm} \in \mathbf{P}\lambda_{\pm}.} \tag{3.19}$$

Therefore $\kappa_{\pm}$ may be expressed in the form

$$\kappa_{\pm} = (u_{\pm} + v_{\pm}q_1)\lambda_{\pm}, \tag{3.20}$$

$u_\pm$ and $v_\pm$ complex ($\vec{x}$ in the form 3.5, see 3.7). From 3.17 we obtain

$$\begin{aligned} \kappa_\pm^\dagger \kappa_\pm &= (u_\pm^* u_\pm + v_\pm^* v_\pm)\lambda_\pm = \lambda_\pm, \\ \kappa_\pm^\dagger \kappa_\mp &= (u_\pm^* v_\mp - v_\pm^* u_\mp) q_1 \lambda_\mp = 0, \end{aligned} \tag{3.21}$$

implying

$$\begin{aligned} u_\pm^* u_\pm + v_\pm^* v_\pm &= 1, \\ u_\pm^* v_\mp - v_\pm^* u_\mp &= 0. \end{aligned} \tag{3.22}$$

The symmetry of the equations 3.22 is $U_+(1) \times U_-(1) \times SU(2)$. The $SU(2)$ arises from the left action of the q_j. Relative to this action u_+ and v_+ transform as the components of an $SU(2)$ doublet, as do u_- and v_- (see 3.7). $U_\pm(1)$ arises from the action of $i\lambda_\pm$ from the right. In particular note that

$$\begin{aligned} e^{i\lambda_+\theta} &= 1 + i\lambda_+\theta + \tfrac{1}{2}(i\lambda_+\theta)^2 + ... \\ &= \lambda_- + \lambda_+ + \lambda_+[i\theta] + \lambda_+[\tfrac{1}{2}(i\theta)^2] + ... \\ &= \lambda_- + \lambda_+[1 + i\theta + \tfrac{1}{2}(i\theta)^2 + ...] \\ &= \lambda_- + \lambda_+[e^{i\theta}]. \end{aligned}$$

Therefore, for example,

$$\kappa_+ e^{i\lambda_+\theta} = \kappa_+(\lambda_- + \lambda_+ e^{i\theta}) = \kappa_+ e^{i\theta},$$

$$\kappa_- e^{i\lambda_+\theta} = \kappa_-(\lambda_- + \lambda_+ e^{i\theta}) = \kappa_-,$$

so $i\lambda_\pm$ from the right acts nontrivially only on $\kappa_\pm$.

We may re-express all this in a more conventional way using matrices, which is done to help the reader get a handle on the mathematics before we get to the more complicated triple tensor product algebra of the next section. The $\lambda_\pm$ can be represented as follows:

$$\lambda_+ \rightarrow \begin{bmatrix} 1 & 0 \\ 0 & 0 \end{bmatrix}, \lambda_- \rightarrow \begin{bmatrix} 0 & 0 \\ 0 & 1 \end{bmatrix},$$

so that

$$\kappa_+ \rightarrow \begin{bmatrix} u_+ & 0 \\ v_+ & 0 \end{bmatrix}, \kappa_- \rightarrow \begin{bmatrix} 0 & u_- \\ 0 & v_- \end{bmatrix}.$$

$\kappa_\pm^\dagger \kappa_\mp = 0$ implies

$$\begin{bmatrix} 0 & u_- \\ 0 & v_- \end{bmatrix} = e^{i\theta} \begin{bmatrix} 0 & -v_+^* \\ 0 & u_+^* \end{bmatrix}.$$

In this case the left action of a 2×2 representation of $U(1) \times SU(2)$ on λ_+ accounts for $\kappa_+ = \begin{bmatrix} u_+ & 0 \\ v_+ & 0 \end{bmatrix}$, and the map M_1 takes this to $\begin{bmatrix} 0 & -v_+^* \\ 0 & u_+^* \end{bmatrix}$. Finally the $U(1)$ phase $e^{i\theta}$ completes the action of $[U(1) \times SU(2)] \times U(1)$. Schematically

$$\boxed{\begin{array}{llll} \lambda_+ & [U(1) \times SU(2)] & & \rightarrow \kappa_+ \\ \lambda_- & [U(1) \times SU(2)] & \times U_-(1) & \rightarrow \kappa_- \end{array}} \tag{3.23}$$

We may now re-express 3.16 in the general form

$$\boxed{< A, B >= \tfrac{1}{4}\{[\textstyle\sum_{rs} M_r((A\kappa_s)^\dagger (B\kappa_s))] + [same]^\dagger\}.} \tag{3.24}$$

3.3. Tensoring Three: Clifford Algebras and Spinors.

Let

$$\boxed{\mathbf{T} = \mathbf{C} \otimes \mathbf{Q} \otimes \mathbf{O},}$$

and

$$\boxed{\mathbf{T}_L = \mathbf{C}_L \otimes \mathbf{Q}_L \otimes \mathbf{O}_L = \mathbf{P}_L \otimes \mathbf{O}_L.}$$

As $\mathbf{P}_L$ is viewed as the Pauli algebra of (1,3)-space-time, from which the Dirac algebra for that space is built, so too will we view $\mathbf{T}_L$ as the "Pauli" algebra of (1,9)-space-time, from which the "Dirac" algebra of that space will be built. In particular note that since $\mathbf{P}_L \simeq \mathbf{C}(2)$, and $\mathbf{O}_L \simeq \mathbf{R}(8)$, then

$$\boxed{\mathbf{T}_L \simeq \mathbf{C}(16),}$$

and so we may identify it with the Clifford algebra $\mathbf{R}_{0,9}$.

The object space of $\mathbf{R}_{0,9} \simeq \mathbf{C}(16)$ is the space of 16×1 complex spinors. As was the case for $\mathbf{P}$ and $\mathbf{P}_L$, the $\mathbf{P}$ idempotents $\lambda_\pm$ defined in 3.1 resolve the elements X of $\mathbf{T}$ (which is 32-dimensional over $\mathbf{C}$) into $X\lambda_+$ and $X\lambda_-$, a pair of 16-dimensional (over $\mathbf{C}$) $\mathbf{R}_{0,9} \simeq \mathbf{T}_L$ spinors. And again, the $SU(2)$ action 3.5 on X in $\mathbf{T}$ commutes with $\mathbf{T}_L$ and is therefore internal. The example action developed in 3.6-8 goes through without change.

It should seem surprising that $\mathbf{T}$ relative to $\mathbf{T}_L$ has the same internal structure as $\mathbf{P}$ relative to $\mathbf{P}_L$. In both cases the extra internal structure arises from $\mathbf{Q}_R$, which commutes with $\mathbf{Q}_L$, hence also with the external geometric algebras $\mathbf{P}_L$ and $\mathbf{T}_L$. No internal depth arises from the $\mathbf{C}$ because $i \in \mathbf{T}_L$ and is identified below as a basis for the 9-vector subspace of $\mathbf{R}_{0,9}$ (when we expand to Dirac algebras, i will regain its independent, nongeometric status). Also, although $\mathbf{O}$ is noncommutative, never-the-less $\mathbf{O}_L = \mathbf{O}_R$ (see chapter 2). Hence the parts of $\mathbf{T}_R$, or $\mathbf{P}_R$, distinct from and commuting with $\mathbf{T}_L$, or $\mathbf{P}_L$, are all associated with $\mathbf{Q}_R$, and so the internal structures of $\mathbf{T}$ and $\mathbf{P}$ are the same.

The chosen multivector bases for $\mathbf{T}_L \simeq \mathbf{R}_{0,9}$ are ($j \in \{1,2,3\}$, $p,q,r \in \{1,...,6\}$:

<u>*Clifford Algebra* $\mathbf{T}_L \simeq \mathbf{R}_{0,9}$</u>

scalar	1	
vector	$ie_{L7}q_j,\, e_{Lp}$	
2-vector	$q_j,\, e_{Lpq},\, iq_je_{Lp7}$	
3-vector	$ie_{L7},\, iq_je_{Lpq7},\, q_je_{Lp},\, e_{Lpqr}$	
4-vector	$q_je_{Lpq},\, e_{Lpq7},\, ie_{Lp7},\, iq_je_{Lpqr}$	(3.25)
5-vector	$iq_je_{Lpq},\, ie_{Lpq7},\, e_{Lp7},\, q_je_{Lpqr}$	
6-vector	$e_{L7},\, q_je_{Lpq7},\, iq_je_{Lp},\, ie_{Lpqr}$	
7-vector	$iq_j,\, ie_{Lpq},\, q_je_{Lp7}$	
8-vector	$e_{L7}q_j,\, ie_{Lp}$	
9-vector	$i.$	

As usual the space of 2-vectors closes under commutation, and is in this case isomorphic to

$$so(9) \simeq spin(9).$$

The subset $\{q_j\}$ is a basis for

$$so(3) \simeq su(2) \simeq spin(3),$$

and the subset $\{e_{Lpq}\}$ a basis for

$$so(6) \simeq su(4) \simeq spin(6).$$

As has been pointed out, the intersection of this $so(6)$ with LG_2 is $su(3)$ (see 2.58). This $su(3)$ generates the color symmetry in what follows, hence color $SU(3)$ is internal only relative to $\mathbf{R}^{1,3}$, not to $\mathbf{R}^{1,9}$, and in this latter case $SU(3)$ rotations are space rotations (see chapter 6).

3.4. Tensoring Three: Spinor Inner Product.

If A and B are elements of $\mathbf{T}$, then $A\lambda_\pm$ and $B\lambda_\pm$ are four $\mathbf{R}_{0,9}$ spinors. Put into the righthand side of 3.16, the result of this algebraically more complicated case is no longer real, but is linear in the identity, and the 21 elements, $q_j e_a$, j=1,2,3, a=1,...,7. No automorphism or antiautomorphism will simultaneously change the signs of all these 21 elements, so the introduction of another conjugation (beyond M_1 and Hermitian conjugation) into 3.16 will not reduce the righthand side to a real number.

However, the expression 3.16 for the inner product is constructed from a resolution of the identity of $\mathbf{P}$, and a group of automorphisms permuting the idempotents of the resolution. The same technique may be successfully applied to $\mathbf{T}$, although some care is required, for $\mathbf{T}$ is not only nonassociative (inherited from $\mathbf{O}$), but, as will be shown below, also nonalternative.

RESOLVING THE IDENTITY OF T

Not surprisingly the structure of $\mathbf{T} = \mathbf{C} \otimes \mathbf{Q} \otimes \mathbf{O}$ is more complicated than that of $\mathbf{P}$. $\mathbf{T}$ admits at least three inequivalent resolutions of its identity into 4 orthogonal primitive idempotents. For example, let

$$\begin{aligned} \alpha_0 &= \tfrac{1}{4}(1 + q_1 e_1 + q_2 e_2 + q_3 e_6), \\ \alpha_i &= -q_i \alpha_0 q_i, \; i = 1,2,3 \end{aligned} \tag{3.26}$$

(recall that $\{e_1, e_2, e_6\}$ are a quaternion triple of $\mathbf{O}$). These satisfy

$$\begin{aligned} \alpha_m \alpha_n &= \delta_{mn} \alpha_n, \; m,n \in \{0,1,2,3\}, \\ \alpha_0 + \alpha_1 + \alpha_2 + \alpha_3 &= 1. \end{aligned} \tag{3.27}$$

Note, however, that

$$\alpha_0(\alpha_0 e_7) = \alpha_0(e_7 - q_1 e_5 - q_2 e_3 + q_3 e_4)/4 = e_7/4,$$

$$(\alpha_0^2)e_7 = \alpha_0 e_7 = (e_7 - q_1 e_5 - q_2 e_3 + q_3 e_4)/4.$$

This curious result ($\alpha_0(\alpha_0 e_7) \neq (\alpha_0^2)e_7$) demonstrates that **T** is not an alternative algebra (**O** by itself is). Furthermore it demonstrates that while the α_m may be idempotents, they are not projection operators.

This suggests we restrict the definition of a resolution of the identity to consist of orthogonal primitive projection operators. (In fact this is what is generally required of an idempotent, but so rarely does the issue of non-alternativity arise that it is seldom necessary explicitly to specify this.) In particular, $\Delta_m, m = 0, 1, 2, 3$, will be considered to resolve the identity of **T** if for all X in **T**,

$$\Delta_m(\Delta_n X) = (\Delta_m \Delta_n)X = \delta_{mn}\Delta_n X,$$

$$\Delta_0 + \Delta_1 + \Delta_2 + \Delta_3 = 1. \tag{3.28}$$

In particular, define

$$\boxed{\begin{aligned} \Delta_0 &= (1 + i\vec{x})(1 + ie_7)/4 = \lambda_0 \rho_+, \\ \Delta_1 &= (1 - i\vec{x})(1 + ie_7)/4 = \lambda_1 \rho_+, \\ \Delta_2 &= (1 + i\vec{y})(1 - ie_7)/4 = \lambda_2 \rho_-, \\ \Delta_3 &= (1 - i\vec{y})(1 - ie_7)/4 = \lambda_3 \rho_-, \end{aligned}} \tag{3.29}$$

where $\vec{x} = \sum x^i q_i$, $\vec{y} = \sum y^i q_i$ may be distinct, each satisfying

$$\vec{x}^2 = \vec{y}^2 = -1,$$

and $\rho_\pm = (1 \pm ie_7)/2$, which alone resolve the identity of $\mathbf{C} \otimes \mathbf{O}$.

The $\vec{x}$ into $\vec{y}$ flexibility in the idempotents Δ_m is allowed because $\rho_+\rho_- = 0$ ensures that Δ_0 and Δ_1 are orthogonal to Δ_2 and Δ_3. If we set $\vec{x} = \vec{y}$, then an additional degree of freedom can arise in the octonion. For example, the four idempotents

$$\begin{aligned}\beta_0 &= (1 + i\vec{x})(1 + ie_7)/4,\\ \beta_1 &= (1 - i\vec{x})(1 + ie_5)/4,\\ \beta_2 &= (1 + i\vec{x})(1 - ie_7)/4,\\ \beta_3 &= (1 - i\vec{x})(1 - ie_5)/4,\end{aligned} \tag{3.30}$$

also resolve the identity of **T**, and they satisfy 3.28. But the Δ_m also satisfy

$$\Delta_m(X\Delta_n) = (\Delta_m X)\Delta_n, \tag{3.31}$$

which the β_m do not. So components of X in **T** with respect to the Δ_m, namely $\Delta_m X \Delta_n$, are consistently definable in this case, but in neither the α_m nor β_m cases.

THE TRACE OF X

For all X in **T**,

$$\Delta_m X \Delta_m = X_m \Delta_m, \tag{3.32}$$

where $X_m, m = 0, 1, 2, 3$, are complex. Again we define

$$tr(X) = \sum_m X_m. \tag{3.33}$$

[It is easy to prove 3.32. For $p = 1, ..., 6$,

$$\rho_\pm e_p \rho_\pm = e_p \rho_\mp \rho_\pm = 0$$

(parentheses unnecessary). Also,

$$\rho_\pm ie_7 \rho_\pm = \pm\rho_\pm.$$

Any element X of **T** may be written as

$$X = X^0 + \sum_{a=1}^{7} X^a ie_a, \quad X^0, X^a \in \mathbf{C} \otimes \mathbf{Q}, \; a = 1, ..., 7.$$

Therefore, for example,

$$\begin{aligned}\Delta_0 X \Delta_0 &= \lambda_0 \rho_+ X \rho_+ \lambda_0 \\ &= \lambda_0 (X^0 + X^7) \lambda_0 \rho_+ .\end{aligned}$$

We have already seen that for elements $Y \in \mathbf{C} \otimes \mathbf{Q}$,

$$\lambda_0 Y \lambda_0 = Y_0 \lambda_0,$$

$Y_0 \in \mathbf{C}$. Therefore,

$$\lambda_0 (X^0 + X^7) \lambda_0 \rho_+ = X_0 \lambda_0 \rho_+ = X_0 \Delta_0,$$

$X_0 \in \mathbf{C}$. (This confusing use of indices is now at an end.)

Let $M_m, m = 0, 1, 2, 3$, be a group of automorphisms on $\mathbf{T}$ defined by

$$\boxed{\begin{aligned} M_0(X) &= X, \\ M_1(X) &= I_7[X^*], \\ M_2(X) &= -q_2 I_7[X] q_2, \\ M_3(X) &= -q_2 X^* q_2 \end{aligned}} \tag{3.34}$$

(I_7 could have been replaced with $I_p, p = 3, 5, 6$), where

$$X \to X^*$$

is complex conjugation, and

$$X \to -q_2 X q_2$$

changes the signs of the quaternion units $q_r, r = 1, 3$. I_7 is defined in chapter 2 (if $p = 3, 5, 6, 7$, then $I_p(e_7) = -e_7$, and in particular

$$I_7(e_p) = -e_p, \; p = 1, 2, 4, 7).$$

We may assume without a loss in generality that

$$\vec{y} = -q_2 \vec{x} q_2.$$

Therefore,

$$M_m(\Delta_0) = \Delta_m \tag{3.35}$$

(in addition, $M_1 M_2 = M_3$, $M_m^2 = M_0$ for all m). The actions of the automorphisms M_m, and the M_m followed by Hermitian conjugation, on **T** are given below ($r = 1, 2$, $p = 1, 2, 4, 7$, $q = 3, 5, 6$):

(3.36)

	M_0	M_1	M_2	M_3	$M_0^\dagger$	$M_1^\dagger$	$M_2^\dagger$	$M_3^\dagger$
1	+	+	+	+	+	+	+	+
i	+	−	+	−	−	+	−	+
q_r	+	+	−	−	−	−	+	+
q_2	+	+	+	+	−	−	−	−
e_p	+	−	−	+	−	+	+	−
e_q	+	+	+	+	−	−	−	−
iq_r	+	−	−	+	+	−	−	+
iq_2	+	−	+	−	+	−	+	−
ie_p	+	+	−	−	+	+	−	−
ie_q	+	−	+	−	+	−	+	−
$q_r e_p$	+	−	+	−	+	−	+	−
$q_2 e_p$	+	−	−	+	+	−	−	+
$q_r e_q$	+	+	−	−	+	+	−	−
$q_2 e_q$	+	+	+	+	+	+	+	+
$iq_r e_p$	+	+	+	+	−	−	−	−
$iq_2 e_p$	+	+	−	−	−	−	+	+
$iq_r e_q$	+	−	−	+	−	+	+	−
$iq_2 e_q$	+	−	+	−	−	+	−	+

Therefore,

$$X_0 + X_1 + X_2 + X_3 = \sum_m \Delta_m [X + M_1(X^\dagger) + M_2(X) + M_3(X^\dagger)]\Delta_m. \tag{3.37}$$

The term in the brackets of 3.37 is 4 times those parts of X linear in $1, i, q_2 e_p$, and $iq_2 e_p, p = 3, 5, 6$. But $\Delta_m e_p \Delta_m = 0$, $p = 3, 5, 6$ (because $\rho_\pm e_p \rho_\pm = \rho_\pm \rho_\mp e_p = 0$), therefore $\sum X_m$ is just four times the complex part of X.

Using 3.37 we can construct an inner product of A and B in **T**, which is just the real part of $A^\dagger B$. We employ an expression for the inner product identical in form to 3.16:

$$\boxed{\begin{aligned} < A, B > &= \tfrac{1}{8}(tr(A^\dagger B + B^\dagger A)) \\ &= \tfrac{1}{8}\{[\textstyle\sum_{mn} M_m((A\Delta_n)^\dagger (B\Delta_n))] + [same]^\dagger\}. \end{aligned}} \tag{3.38}$$

3.5. Derivation of the Standard Symmetry.

As was done in 3.24 relative to 3.16, the expression 3.38 for $< A, B >$ can be generalized by replacing Δ_n in 3.38 by the algebraically more complicated $\Gamma_n, n = 0, 1, 2, 3$:

$$\boxed{< A, B >= \tfrac{1}{8}\{[\textstyle\sum_{mn} M_m((A\Gamma_n)^\dagger(B\Gamma_n))] + [same]^\dagger\}.} \tag{3.39}$$

We require that with respect to the Γ_n the components of the identity be the same as those obtained from Δ_n. That is,

$$\boxed{\Gamma_n^\dagger \Gamma_m = \delta_{nm}\Delta_m} \tag{3.40}$$

(see 3.17). In addition we need a consistency requirement, that the components of any X in $\mathbf{T}$, with respect to the Γ_n, be free of nonassociativity problems. That is,

$$(\Gamma_n^\dagger X)\Gamma_m = \Gamma_n^\dagger(X\Gamma_m), n, m \in \{0, 1, 2, 3\}. \tag{3.41}$$

From 3.40 we obtain

$$\Gamma_n^\dagger \Gamma_n = \Delta_n,$$

so

$$\Delta_m(\Gamma_n^\dagger \Gamma_n)\Delta_m = (\Gamma_n\Delta_m)^\dagger(\Gamma_n\Delta_m) = \delta_{mn}\Delta_n,$$

(making use of the fact that $\rho_\pm(AB)\rho_\pm = (\rho_\pm A)(B\rho_\pm)$, and $\Delta_n^\dagger = \Delta_n$). But if $A \in \mathbf{T}$, and $A^\dagger A = 0$, then $A = 0$. Therefore,

$$\Gamma_n\Delta_m = 0 \text{ if } n \neq m.$$

So

$$\Gamma_n = \Gamma_n(\Delta_0 + \Delta_1 + \Delta_2 + \Delta_3) = \Gamma_n\Delta_n.$$

That is,

$$\boxed{\Gamma_n \in \mathbf{T}\Delta_n.} \tag{3.42}$$

For the sake of simplicity, and without loss of generality, we now put the 2-dimensional degree of freedom in $\vec{x}$ and $\vec{y}$ in Δ_m (see 3.29) in the (q_2, q_3)-plane. That is, set

$$\vec{x} = x^2 q_2 + x^3 q_3,$$
$$\vec{y} = -q_2 \vec{x} q_2 = x^2 q_2 - x^3 q_3.$$

So $\{q_1, \vec{x}, q_1\vec{x}\}$ and $\{q_1, \vec{y}, q_1\vec{y}\}$ are two orthonormal bases for the hypercomplex parts of **Q**. Therefore, since

$$\vec{x}(1 \pm i\vec{x})/2 = \mp i(1 \pm i\vec{x})/2,$$
$$\vec{y}(1 \pm i\vec{y})/2 = \mp i(1 \pm i\vec{y})/2,$$
$$e_7 \rho_\pm = \mp i \rho_\pm, \tag{3.43}$$

and $e_3 = -e_2 e_7$, $e_5 = -e_1 e_7$, and $e_6 = -e_4 e_7$, the elements Γ_n of $\mathbf{T}\Delta_n$ can be written in the reduced form

$$\Gamma_n = \omega_n \Delta_n = (u_n^0 + v_n^0 q_1 + (u_n^1 e_1 + u_n^2 e_2 + u_n^4 e_4) + (v_n^1 e_1 + v_n^2 e_2 + v_n^4 e_4) q_1) \Delta_n$$
$$= (u_n^0 + v_n^0 q_1 + \overleftarrow{u}_n + \overleftarrow{v}_n q_1) \Delta_n, \tag{3.44}$$

where the u, v coefficients are complex.

This can be further simplified by imposing the associativity condition 3.41 at this point. Suppose $\overleftarrow{u}_n \neq 0$ and/or $\overleftarrow{v}_n \neq 0$. For example, let $\Gamma_0 = e_1 \Delta_0 = e_1 \lambda_0 \rho_+$. Consider in this case the (0,0) component of e_2 (note: $(\rho_+ e_1) e_2 = \rho_-(e_1 e_2)$):

$$\begin{aligned}
(\Gamma_0^\dagger e_2)\Gamma_0 &= -[(\lambda_0 \rho_+ e_1) e_2](e_1 \lambda_0 \rho_+) \\
&= -[\lambda_0 \rho_-(e_1 e_2)](e_1 \lambda_0 \rho_+) \\
&= -(\rho_- e_6)(e_1 \rho_+)\lambda_0 \\
&= -\rho_+(e_6(e_1 \rho_+))\lambda_0 \\
&= -\rho_+((e_6 e_1)\rho_-)\lambda_0 \\
&= -\lambda_0 \rho_+ e_2 \rho_- \\
&= -\Delta_0 e_2.
\end{aligned}$$

It's bad enough that this diagonal component is nonzero (it should pick out the complex (**C**) part of e_2, which of course is zero), but in addition,

$$\Gamma_0^\dagger(e_2\Gamma_0) = -e_2\Delta_0 \neq -\Delta_0 e_2 = (\Gamma_0^\dagger e_2)\Gamma_0.$$

Avoidance of these problems is possible only if

$$\overleftarrow{u}_n = 0, \overleftarrow{v}_n = 0, \; n = 0,1,2,3,$$

that is, if

$$\Gamma_n = (u_n + v_n q_1)\Delta_n, \tag{3.45}$$

n=0,1,2,3, u_n and v_n complex. In this case, because $\rho_\pm\rho_\mp = 0$, it follows automatically that $\Gamma_n^\dagger\Gamma_m = 0$ if n=0,1 (or 2,3), and m=2,3 (or 0,1). Therefore the only nontrivial equations left in 3.40 are

$$\Gamma_0^\dagger\Gamma_1 = 0,$$

$$\Gamma_2^\dagger\Gamma_3 = 0, \tag{3.46}$$

and their Hermitian conjugates. But because of 3.45, the symmetry of

$$\Gamma_m^\dagger\Gamma_m = \Delta_m, m = 0,1,$$

$$\Gamma_0^\dagger\Gamma_1 = 0,$$

is the same as the symmetry of 3.17, that is,

$$U_0(1) \times U_1(1) \times SU_+(2).$$

The same is true of the equations

$$\Gamma_m^\dagger\Gamma_m = \Delta_m, m = 2,3,$$

$$\Gamma_2^\dagger\Gamma_3 = 0,$$

the

$$U_2(1) \times U_3(1) \times SU_-(2)$$

symmetry of this case being independent of that of the (0,1)-case. So the total symmetry to this point is

$$\boxed{U_0(1) \times U_1(1) \times SU_+(2) \times U_2(1) \times U_3(1) \times SU_-(2)} \tag{3.47}$$

('+' associated with m=0,1, '−' with m=2,3).

The $SU(2)$'s arise from the $q_j, j = 1,2,3$, and the $U(1)$'s from the complex unit i, or the octonion unit e_7. In particular, if $\vec{\alpha}^{\pm} = \sum_1^3 \alpha^{j\pm} q_j$, $\alpha^{j\pm}$ real, and θ^m real, m=0,1,2,3, then

$$\begin{aligned} \Gamma_n &\rightarrow exp[\vec{\alpha}^+ \rho_+ + \vec{\alpha}^- \rho_-]\Gamma_n exp[\textstyle\sum_0^3 i\theta^m \Delta_m] \\ &= exp[\vec{\alpha}^+ \rho_+]exp[\vec{\alpha}^- \rho_-]\Gamma_n \textstyle\prod_0^3 exp[i\theta^m \Delta_m] \end{aligned} \tag{3.48}$$

performs the rotation 3.47. For example,

$$\begin{aligned} exp(\vec{\alpha}^- \rho_-)\Gamma_n &= (\rho_+ + \rho_- exp(\vec{\alpha}^-))\Gamma_n \\ &= exp(\vec{\alpha}^-)\Gamma_n,\ n = 2,3, \\ &= \Gamma_n,\ n = 0,1. \end{aligned} \tag{3.49}$$

The symmetry 3.47 of 3.40 and 3.41 is not complete. The $SU(3)$ stability subgroup of G_2 leaving e_7 fixed (see 2.57) also clearly maintains the validity of 3.40 and 3.41, for the Δ_n and Γ_n are invariant under this $SU(3)$. (Since $SU(3)$ is a subgroup of the automorphism group G_2, it may be extended to the spinors A and B as well.) So each of the Δ_m is subject to a symmetry

$$\boxed{U(1) \times SU(2) \times SU(3)} \tag{3.50}$$

(the $SU(3)$'s may be considered independent for m=0,1,2,3, but as all the Γ_m's are invariant, this is irrelevant).

In 3.48 one method of generating the $SU(2)$ actions was presented. There is another way that will prove more useful in applications. First redefine

$$\boxed{\lambda_{\pm} = \tfrac{1}{2}(1 \pm i(x^2 q_2 + x^3 ie_7 q_3)) = \tfrac{1}{2}(1 \pm i\tilde{x})} \tag{3.51}$$

($\tilde{x}^2 = -1$), so that

$$\boxed{\begin{aligned} \lambda_+\rho_+ &= \lambda_0\rho_+ = \Delta_0, \\ \lambda_-\rho_+ &= \lambda_1\rho_+ = \Delta_1, \\ \lambda_+\rho_- &= \lambda_2\rho_- = \Delta_2, \\ \lambda_-\rho_- &= \lambda_3\rho_- = \Delta_3 \end{aligned}} \tag{3.52}$$

(note: $ie_7\rho_{\pm} = \pm\rho_{\pm}$).

Define

$$\tilde{\alpha} = \alpha^1 ie_7 q_1 + \alpha^2 q_2 + \alpha^3 ie_7 q_3 \tag{3.53}$$

(note that the real span of the set $\{ie_7q_1, q_2, ie_7q_3\}$ closes under commutation and is isomorphic to $su(2)$). Let

$$\vec{\alpha} = \alpha^1 q_1 + \alpha^2 q_2 + \alpha^3 q_3. \tag{3.54}$$

Then

$$\begin{aligned} e^{\tilde{\alpha}}\Delta_0 &= e^{\vec{\alpha}}\Delta_0 &= M_0(e^{\tilde{\alpha}}\Delta_0), \\ e^{\tilde{\alpha}}\Delta_1 &= e^{\vec{\alpha}}\Delta_1 &= M_1(e^{\tilde{\alpha}}\Delta_0), \\ e^{\tilde{\alpha}}\Delta_2 &= e^{-q_2\vec{\alpha}q_2}\Delta_2 &= M_2(e^{\tilde{\alpha}}\Delta_0), \\ e^{\tilde{\alpha}}\Delta_3 &= e^{-q_2\vec{\alpha}q_2}\Delta_3 &= M_3(e^{\tilde{\alpha}}\Delta_0) \end{aligned} \tag{3.55}$$

(note that $M_n(\tilde{\alpha}) = \tilde{\alpha}$). That is, $\tilde{\alpha}$ generates an $SU(2)$ that is carried intact by the M_n from Δ_0 to the other Δ_n. (There is an additional $SU(2)$ degree of freedom on Δ_2 and Δ_3 (see 3.47 and 3.48). We can account for this by an $SU(2)$ action like 3.49 that acts nontrivially only on Δ_2 and Δ_3.)

For each n=0,1,2,3, Δ_n and $q_1\Delta_n$ form an $SU(2)$ doublet with respect to $e^{\tilde{\alpha}}$. The diagonal $SU(2)$ generator is $\tilde{x}$. In particular,

$$\begin{aligned} \tilde{x}\begin{bmatrix} \Delta_n \\ q_1\Delta_n \end{bmatrix} &= \begin{bmatrix} \tilde{x}\Delta_n \\ -q_1\tilde{x}\Delta_n \end{bmatrix} \\ &= \begin{bmatrix} -i & 0 \\ 0 & i \end{bmatrix}\begin{bmatrix} \Delta_n \\ q_1\Delta_n \end{bmatrix}, \text{ n=0,2,} \\ &= \begin{bmatrix} i & 0 \\ 0 & -i \end{bmatrix}\begin{bmatrix} \Delta_n \\ q_1\Delta_n \end{bmatrix}, \text{ n=1,3} \end{aligned}$$

(see 3.52).

The standard symmetry is then identified with a single copy of $U(1) \times SU(2) \times SU(3)$ that is associated with Δ_0 and Γ_0, and carried from there by the M_n to the remaining Δ_n and Γ_n. The other symmetry degrees of freedom that arise on the Δ_j, $j = 1,2,3$, then arise separately. In particular, the symmetry 3.47, together with $SU(3)$, can be rewritten

$$\boxed{[U(1) \times SU(2) \times SU(3)] \times U_1(1) \times U_2(1) \times U_3(1) \times SU_{23}(2),} \tag{3.56}$$

where $[U(1) \times SU(2) \times SU(3)]$ is shared by all the Γ_n (carried along by the maps M_n), $U_j(1), j = 1,2,3$, acts only on Γ_j, and $SU_{23}(2)$ acts only on Γ_2

and Γ_3. Note that the $SU_-(2)$ in 3.47 is a mix of $SU_{23}(2)$ and the $SU(2)$ in brackets. Schematically,

Δ_0	$[U(1) \times SU(2) \times SU(3)]$			$\rightarrow \Gamma_0$
Δ_1	$[U(1) \times SU(2) \times SU(3)]$	$\times U_1(1)$		$\rightarrow \Gamma_1$
Δ_2	$[U(1) \times SU(2) \times SU(3)]$	$\times U_2(1)$	$\times SU_{23}(2)$	$\rightarrow \Gamma_2$
Δ_3	$[U(1) \times SU(2) \times SU(3)]$	$\times U_3(1)$	$\times SU_{23}(2)$	$\rightarrow \Gamma_3$.

[Curiously $U(1)^4 \times SU(2)^2 \times SU(3)$ also arises in **[11]**, where the division algebras are applied to the problem of particle masses. It is not yet known if there is a connection.]

3.6. $SU(2) \times SU(3)$ Multiplets, and $U(1)$.

Let $X = X^0 + \sum_1^7 X^a e_a$ be an element of **O**.

$$\begin{aligned} \rho_+ X &= \rho_+[(X^0 - iX^7) + (X^1 - iX^5)e_1 + (X^2 - iX^3)e_2 + (X^4 - iX^6)e_4] \\ &= \rho_+[Z^0 + Z^1 e_1 + Z^2 e_2 + Z^4 e_4]. \end{aligned} \tag{3.57}$$

The $\rho_+ Z^0$ part of 3.57 transforms as a singlet with respect to $SU(3)$. The rest of $\rho_+ X$ transforms as a triplet. This may be seen most explicitly by representing a basis for $su(3)$ in $\mathbf{O}_L$ adjointly as 3x3 matrices acting on the column $\begin{bmatrix} Z^1 \\ Z^2 \\ Z^4 \end{bmatrix}$. This is done below:

$$\begin{aligned}
&\tfrac{1}{2}(e_{L34}-e_{L26}) \to \begin{bmatrix} \cdot & \cdot & \cdot \\ \cdot & \cdot & i \\ \cdot & i & \cdot \end{bmatrix}, \tfrac{1}{2}(e_{L63}-e_{L24}) \to \begin{bmatrix} \cdot & \cdot & \cdot \\ \cdot & \cdot & 1 \\ \cdot & -1 & \cdot \end{bmatrix}, \\
&\tfrac{1}{2}(e_{L61}-e_{L45}) \to \begin{bmatrix} \cdot & \cdot & i \\ \cdot & \cdot & \cdot \\ i & \cdot & \cdot \end{bmatrix}, \tfrac{1}{2}(e_{L56}-e_{L41}) \to \begin{bmatrix} \cdot & \cdot & 1 \\ \cdot & \cdot & \cdot \\ -1 & \cdot & \cdot \end{bmatrix}, \\
&\tfrac{1}{2}(e_{L52}-e_{L13}) \to \begin{bmatrix} \cdot & i & \cdot \\ i & \cdot & \cdot \\ \cdot & \cdot & \cdot \end{bmatrix}, \tfrac{1}{2}(e_{L35}-e_{L12}) \to \begin{bmatrix} \cdot & 1 & \cdot \\ -1 & \cdot & \cdot \\ \cdot & \cdot & \cdot \end{bmatrix}, \\
&\tfrac{1}{2}(e_{L46}-e_{L23}) \to \begin{bmatrix} \cdot & \cdot & \cdot \\ \cdot & i & \cdot \\ \cdot & \cdot & -i \end{bmatrix}, \tfrac{1}{2}(e_{L23}-e_{L15}) \to \begin{bmatrix} i & \cdot & \cdot \\ \cdot & -i & \cdot \\ \cdot & \cdot & \cdot \end{bmatrix}.
\end{aligned} \tag{3.58}$$

For example,

$$\tfrac{1}{2}(e_{L34}-e_{L26})(\rho_+e_1) = \tfrac{1}{2}\rho_+[e_3(e_4e_1) - e_2(e_6e_1)] = 0,$$

$$\tfrac{1}{2}(e_{L34}-e_{L26})(\rho_+e_2) = \tfrac{1}{2}\rho_+[e_3(e_4e_2) - e_2(e_6e_2)] = \tfrac{1}{2}\rho_+(-2e_6) = i\rho_+e_4,$$

$$\tfrac{1}{2}(e_{L34}-e_{L26})(\rho_+e_4) = \tfrac{1}{2}\rho_+[e_3(e_4e_4) - e_2(e_6e_4)] = \tfrac{1}{2}\rho_+(-2e_3) = i\rho_+e_2.$$

Assigning ρ_+X the multiplet structure

$$\mathbf{1} \oplus \mathbf{3}$$

(singlet plus triplet) means ρ_-X transforms as

$$\overline{\mathbf{1}} \oplus \overline{\mathbf{3}}$$

(antisinglet plus antitriplet). At this stage, however, the ρ_+X and ρ_-X multiplets are merely conjugates of each other, hence not independent.

Now let $X \in \mathbf{C} \otimes \mathbf{O}$, so $X^a \in \mathbf{C}, a = 0, 1, ..., 7$, in 3.57. In this case ρ_+X and ρ_-X have independent multiplets (ie., they are no longer conjugates of each other). Therefore, under

$$\boxed{SU(3) : \mathbf{C} \otimes \mathbf{O} \to \mathbf{1} \oplus \mathbf{3} \oplus \overline{\mathbf{1}} \oplus \overline{\mathbf{3}}.} \tag{3.59}$$

Therefore, $\mathbf{T} = (\mathbf{C} \otimes \mathbf{O}) \otimes \mathbf{Q}$ has the same multiplet structure as $\mathbf{C} \otimes \mathbf{O}$, but times four. Explicitly,

$$\boxed{\begin{array}{ll} \rho_+\mathbf{T}\rho_+ \to \mathbf{1}; & \rho_+\mathbf{T}\rho_- \to \mathbf{3}, \\ \rho_-\mathbf{T}\rho_- \to \overline{\mathbf{1}}; & \rho_-\mathbf{T}\rho_+ \to \overline{\mathbf{3}}. \end{array}} \tag{3.60}$$

Suppose in addition that X transforms with respect to $SU(2)$ as $\Gamma_n^\dagger$. That is,

$$X \to Xe^{-\tilde{\alpha}}. \tag{3.61}$$

So X transforms like $SU(2)$ doublets (see 3.9 and 3.56).

[Note: of course we could as well have let X transform with respect to $SU(2)$ as Γ_n, so that $X \to e^{\tilde{\alpha}}X$, but for two reasons 3.61 is preferable. Firstly, with that action the product $X\Gamma_n$ will be $SU(2)$ invariant; and secondly, if $SU(2)$ acts on the spinor X from the right, then it is internal with respect to the action of the geometric algebra $\mathbf{R}_{1,3}$ from the left. This aspect is developed in chapter 4.]

Thus $SU(3)$ divides $\mathbf{T}$ into 8 spinor bits (see 3.59), and $SU(2)$ divides each of these in two, making 16 bits in all. Each of these 16 $SU(2) \times SU(3)$ bits is 64/16=4-dimensional over the reals, 2-dimensional over $\mathbf{C}$. In fact they are Weyl spinors.

Consider for example the $SU(3)$ singlet, $SU(2)$ doublet,

$$\{\rho_+\mathbf{T}\Delta_0,\ \rho_+\mathbf{T}\Delta_1\}.$$

Each of these subspaces is $SU(3)$ invariant, and they are mixed by the action of $e^{-\tilde{\alpha}}$ from the right. They are also stable under the action of $\mathbf{P}_L$, the Pauli algebra, and with respect to that algebra they transform as Pauli spinor spaces.

We conclude, then, that with respect to $SU(2) \times SU(3)$, $\mathbf{T}$ transforms exactly like the direct sum of a family of lefthanded leptons and quarks, and an antifamily of righthanded antileptons and antiquarks.

$U(1)$ CHARGES

As was done in 3.55 for $SU(2)$, we may carry a $U(1)$ charge from Γ_0 to the other Γ_j, j=1,2,3, via the M_j. That is,

$$\begin{aligned}
\Gamma_0 &\rightarrow e^{i\theta}\Gamma_0 &= M_0(e^{i\theta}\Gamma_0) &= M_0(e^{-e_7\theta}\Gamma_0),\\
\Gamma_1 &\rightarrow e^{-i\theta}\Gamma_1 &= M_1(e^{i\theta}\Gamma_0) &= M_1(e^{-e_7\theta}\Gamma_0),\\
\Gamma_2 &\rightarrow e^{i\theta}\Gamma_2 &= M_2(e^{i\theta}\Gamma_0) &= M_2(e^{-e_7\theta}\Gamma_0),\\
\Gamma_3 &\rightarrow e^{-i\theta}\Gamma_3 &= M_3(e^{i\theta}\Gamma_0) &= M_3(e^{-e_7\theta}\Gamma_0).
\end{aligned} \tag{3.62}$$

In 3.55, Γ_1 and Γ_3 transform under a conjugate $SU(2)$ representation relative to Γ_0 and Γ_2 (see 3.56). The same is true of the $U(1)$ in 3.62. Therefore they combine to form

$$U(1) \times SU(2) \rightarrow U(2). \tag{3.63}$$

The remaining three $U(1)$ degrees of freedom can be performed separately on Γ_j, j=1,2,3.

One final note: the $U(1)$ transformation

$$\Gamma_m \rightarrow e^{-\phi e_7/6}\Gamma_m e^{\phi e_7/6} \tag{3.64}$$

(the factor 1/6 sets a conventional scale) acts trivially on the Γ_m, but on the other elements of $\mathbf{T}$ combines with $SU(3)$ to form $U(3)$. Note that if $X \in \mathbf{T}$, then

$$X \rightarrow e^{-\phi e_7/6}Xe^{\phi e_7/6} = \rho_+X\rho_+ + \rho_-X\rho_- + e^{i\phi/3}\rho_+X\rho_- + e^{-i\phi/3}\rho_-X\rho_+. \tag{3.65}$$

So $\rho_+X\rho_+ \rightarrow \mathbf{1}$ and $\rho_-X\rho_- \rightarrow \bar{\mathbf{1}}$ are invariant, and $\rho_+X\rho_- \rightarrow \mathbf{3}$ and $\rho_-X\rho_+ \rightarrow \bar{\mathbf{3}}$ transform as $U(1)$ conjugates. So the total symmetry carried by the M_m from Γ_0 to the other Γ_m, including the trivial $U(1)$ 3.64, is

$$U(2) \times U(3), \tag{3.66}$$

and the quantum numbers arising from this group on the various Pauli (Weyl) spinor components are precisely those of a lefthanded lepto quark family, and righthanded antifamily. Obviously to assume this is coincidental requires a faith and vested interest in some other scheme. To assume it is not coincidental requires a faith and vested interest in this scheme. On the persuasiveness of this latter faith does half of this book rest. As to the other half, while strictly mathematical, it will be seen to bolster the first half.

4. Connecting to Physics.

4.1. Connecting to Geometry.

The parity violation of Dirac spinor fields occurs to the level of the Pauli (or Weyl) subspinor fields. So in physics the fundamental fermion fields are Pauli spinors instead of Dirac spinors. Dirac spinors consist of a pair of Pauli spinors; the Dirac algebra is the Pauli algebra tensored with $\mathbf{R}(2)$. This extra $\mathbf{R}(2)$ is required not only of this physical context, but also of mathematical contexts, like using $\mathbf{O}_L(2)$ to represent $\mathbf{R}_{8,0}$ and triality operations (see chapter 7), and in the construction of the groups of the magic square (see chapter 8). This paradigm, that the Clifford algebra of spacetime starts with a Pauli algebra for just space, and that this when tensored with $\mathbf{R}(2)$ provides room for time, will also be used below in the construction of a Clifford algebra for (1,9)-spacetime.

Define in $\mathbf{R}(2)$,

$$\epsilon = \begin{bmatrix} 1 & 0 \\ 0 & 1 \end{bmatrix}, \alpha = \begin{bmatrix} 1 & 0 \\ 0 & -1 \end{bmatrix},$$

$$\beta = \begin{bmatrix} 0 & 1 \\ 1 & 0 \end{bmatrix}, \omega = \begin{bmatrix} 0 & 1 \\ -1 & 0 \end{bmatrix}. \tag{4.1}$$

These four matrices form a basis for $\mathbf{R}(2)$, and α, β, and ω mutually anticommute.

$\mathbf{P}_L(2) \simeq \mathbf{C}(4)$, the Dirac algebra, which is the complexification of the Clifford algebra $\mathbf{R}_{1,3}$. Therefore a basis for $\mathbf{R}_{1,3}$ can be represented in terms of the elements i and q_{Lj} of $\mathbf{P}_L$, and the matrices in 4.1.

Multivector bases for $\mathbf{R}_{1,3}$ *are* $(j = 1,2,3)$:

$$\begin{array}{ll} \text{scalar} & \epsilon \\ \text{vector} & \beta,\ iq_{Lj}\omega, \\ \text{2-vector} & q_{Lj}\epsilon,\ iq_{Lj}\alpha, \\ \text{3-vector} & i\omega,\ q_{Lj}\beta, \\ \text{4-vector} & i\alpha, \end{array} \tag{4.2}$$

and we define

$$\gamma_0 = \beta,\ \gamma_j = iq_j\omega,\ \gamma_5 = -i\gamma_0\gamma_1\gamma_2\gamma_3 = \alpha. \tag{4.3}$$

The spinor space of $\mathbf{P}_L(2)$ is $\mathbf{P}^2$, 2x1 matrices over $\mathbf{P}$. An element A of $\mathbf{P}^2$ can be decomposed into $A\lambda_r, r = 0,1$, and each of these is a 4-dimensional (complex) Dirac spinor. Left multiplication by the chiral projectors $\frac{1}{2}(\epsilon \pm \gamma_5)$ project a pair of Weyl spinors from each Dirac spinor.

$\mathbf{T}_L(2) \simeq \mathbf{C}(32)$, which is the complexification of $\mathbf{R}_{1,9}$. So $\mathbf{T}_L \simeq \mathbf{R}_{9,0}$ is, so-to-speak, the "Pauli" algebra to the (1,9)-"Dirac" algebra $\mathbf{T}_L(2)$.

[Note: The generalization I am making is an obvious one. $\mathbf{C}_L \otimes \mathbf{Q}_L$ becomes $\mathbf{C}_L \otimes \mathbf{Q}_L \otimes \mathbf{O}_L$. This is a mathematically motivated generalization, for it will be remembered that the principal reason the Clifford algebra of (1,3)-spacetime requires complexification is a physical one, for without complexification we can not construct a chiral projector, which is phenomenologically required. In the Clifford algebra for (1,9)-spacetime, however, no such complexification is required, for chiral projectors can be constructed in the absence of the complex unit i. Hence, mathematics, as opposed to physics, is leading us in what follows. The results will speak for themselves.]

Multivector bases for $\mathbf{R}_{1,9}$ *are* $(j = 1,2,3;\ p,q,r = 1,...,6)$:

scalar	ϵ
vector	$\beta, e_{L7}q_j\omega, ie_{Lp}\omega,$
2-vector	$q_j\epsilon, e_{L7}q_j\alpha, e_{Lpq}\epsilon, ie_{Lp}\alpha, iq_je_{Lp7}\epsilon,$
3-vector	$e_{L7}\omega, q_j\beta, e_{Lpq}\beta, q_je_{Lpq7}\omega, iq_je_{Lp7}\beta, iq_je_{Lp}\omega, ie_{Lpqr}\omega,$
4-vector	$e_{L7}\alpha, q_je_{Lpq}\epsilon, q_je_{Lpq7}\alpha, e_{Lpq7}\epsilon, ie_{Lp7}\epsilon, iq_je_{Lp}\alpha, ie_{Lpqr}\alpha, iq_je_{Lpqr}\epsilon,$
5-vector	$e_{Lpq7}\omega, q_je_{Lpq}\beta, e_{Lpq7}\beta, q_je_{Lpq}\omega, ie_{Lp7}\beta, iq_je_{Lpqr}\omega, ie_{Lp7}\omega, iq_je_{Lpqr}\beta,$
6-vector	$e_{L7}\epsilon, q_je_{Lpq}\alpha, q_je_{Lpq7}\epsilon, e_{Lpq7}\alpha, ie_{Lp7}\alpha, iq_je_{Lp}\epsilon, ie_{Lpqr}\epsilon, iq_je_{Lpqr}\alpha,$
7-vector	$e_{L7}\beta, q_j\omega, e_{Lpq}\omega, q_je_{Lpq7}\beta, iq_je_{Lp7}\omega, iq_je_{Lp}\beta, ie_{Lpqr}\beta,$
8-vector	$q_j\alpha, e_{L7}q_j\epsilon, e_{Lpq}\alpha, ie_{Lp}\epsilon, q_je_{Lp7}\alpha,$
9-vector	$\omega, e_{L7}q_j\beta, ie_{Lp}\beta,$
10-vector	$\alpha.$

(4.4)

Of course, only half of $\mathbf{T}_L(2)$ appears in 4.4 - the rest arises from multiplication by i. Note that with the aid of the pseudovector (10-vector basis) we can in this case construct chiral projectors without complexification. But we are already complexified, and without the inclusion of the algebra $\mathbf{C}$ in $\mathbf{T}$ we could not have constructed a basis for the Clifford algebra $\mathbf{R}_{1,9}$. In this case complexification is mathematically required.

DIMENSIONAL REDUCTION

The object space of $\mathbf{T}_L(2)$ is $\mathbf{T}^2$, 2x1 matrices over $\mathbf{T}$. Each A in $\mathbf{T}^2$ can be decomposed into a pair of complex (1,9)-Dirac spinors, $A\lambda_r, r = 0,1$. $\mathbf{T}^2$ can also be decomposed into $SU(2) \times SU(3)$ family plus antifamily bits using the projectors $\rho_\pm$ and Δ_m (see chapter 3, section 5). In particular,

Family (SU(2) doublets)

$$\{\rho_+\mathbf{T}^2\Delta_0,\ \rho_+\mathbf{T}^2\Delta_1\}\quad \mathrm{SU}(3)\to\mathbf{1},$$

$$\{\rho_+\mathbf{T}^2\Delta_2,\ \rho_+\mathbf{T}^2\Delta_3\}\quad \mathrm{SU}(3)\to\mathbf{3},$$

Antifamily

$$\{\rho_-\mathbf{T}^2\Delta_2,\ \rho_-\mathbf{T}^2\Delta_3\}\quad \mathrm{SU}(3)\to\bar{\mathbf{1}},$$

$$\{\rho_-\mathbf{T}^2\Delta_0,\ \rho_-\mathbf{T}^2\Delta_1\}\quad \mathrm{SU}(3)\to\bar{\mathbf{3}}. \tag{4.5}$$

Therefore, $\rho_+\mathbf{T}^2$ is the family half of $\mathbf{T}^2$, and $\rho_-\mathbf{T}^2$ the antifamily half.

Corresponding to the family projection on the spinor space,

$$\mathbf{T}^2\to\rho_+\mathbf{T}^2,$$

is the projection

$$\boxed{\mathbf{T}_L(2)\to\rho_{L+}\mathbf{T}_L(2)\rho_{L+}} \tag{4.6}$$

on the Clifford algebra, where

$$\rho_{L+}=\frac{1}{2}(1+ie_{L7}).$$

So

$$(\rho_{L+}\mathbf{T}_L(2)\rho_{L+})[\rho_+\mathbf{T}^2]=\rho_+\mathbf{T}^2.$$

$\rho_{L+}\mathbf{T}_L(2)\rho_{L+}$ is a subalgebra of $\mathbf{T}_L(2)$ isomorphic to $\mathbf{C}(16)$. While $\mathbf{C}(16)$ is isomorphic to a Clifford algebra, $\rho_{L+}\mathbf{T}_L(2)\rho_{L+}$ should not be thought of in this way. Rather, from the Clifford algebra $\mathbf{R}_{1,9}\subset\mathbf{T}_L(2)$ is projected the Clifford algebra $\mathbf{R}_{1,3}$ crossed with the 15 generators of $so(6)\simeq spin(6)\simeq su(4)$, and this Lie algebra contains the generators for color $SU(3)$, so they survive the projection, while the extra 6 space dimensions do not.

The (1,9)-multivectors surviving the projection 4.6 are (j=1,2,3; p,q = 1,...,6):

$$
\begin{array}{ll}
 & \underline{\textit{Dimensionally Reduced } \mathbf{R}_{1,9}} \\
\text{scalar} & \boxed{\epsilon\rho_{L+}} \\
\text{vector} & \boxed{\beta\rho_{L+}, iq_j\omega\rho_{L+}} \\
\text{2-vector} & \boxed{q_j\epsilon\rho_{L+}, iq_j\alpha\rho_{L+},}\ e_{Lpq}\epsilon\rho_{L+}, \\
\text{3-vector} & \boxed{i\omega\rho_{L+}, q_j\beta\rho_{L+},}\ e_{Lpq}\beta\rho_{L+}, iq_je_{Lpq}\omega\rho_{L+} \\
\text{4-vector} & \boxed{i\alpha\rho_{L+},}\ iq_je_{Lpq}\alpha\rho_{L+}, q_je_{Lpq}\epsilon\rho_{L+}, ie_{Lpq}\epsilon\rho_{L+} \\
\text{5-vector} & q_je_{Lpq}\beta\rho_{L+}, ie_{Lpq}\beta\rho_{L+}, ie_{Lpq}\omega\rho_{L+}, q_je_{Lpq}\omega\rho_{L+} \\
\text{6-vector} & i\epsilon\rho_{L+}, iq_je_{Lpq}\epsilon\rho_{L+}, q_je_{Lpq}\alpha\rho_{L+}, ie_{Lpq}\alpha\rho_{L+} \\
\text{7-vector} & i\beta\rho_{L+}, q_j\omega\rho_{L+}, e_{Lpq}\omega\rho_{L+}, iq_je_{Lpq}\beta\rho_{L+} \\
\text{8-vector} & q_j\alpha\rho_{L+}, iq_j\epsilon\rho_{L+}, e_{Lpq}\alpha\rho_{L+} \\
\text{9-vector} & \omega\rho_{L+}, iq_j\beta\rho_{L+} \\
\text{10-vector} & \alpha\rho_{L+}
\end{array}
\tag{4.7}
$$

For example, the reduction of the 1-vectors of 4.4 occurs as follows:

$$\rho_{L+}\beta\rho_{L+} = \beta\rho_{L+},$$

$$\rho_{L+}e_{L7}q_j\omega\rho_{L+} = -iq_j\omega\rho_{L+},$$

$$\rho_{L+}ie_{Lp}\omega\rho_{L+} = 0,$$

leaving the vector basis in 4.7, which is a basis for the vector part of $\mathbf{R}_{1,3}\rho_{L+}$ (see 4.2). The rest of this isomorphic copy of $\mathbf{R}_{1,3}\rho_{L+}$ is indicated by boxes in 4.7.

The ten (1,9)-vectors reduce to the four (1,3)-vectors (ρ_{L+} tacked on), but higher multivectors do not similarly reduce simply from (1,9) to (1,3). Under commutation the 2-vectors of $\mathbf{R}_{1,9}$ form the Lie algebra $so(1,9)$. The projection 4.6 reduces this to $so(1,3) \times so(6)$. In all cases, however, the subalgebra 4.7 has a nontrivial effect only on the matter half $\rho_+\mathbf{T}^2$ of $\mathbf{T}^2$.

The antimatter projection

$$\mathbf{T}_L(2) \rightarrow \rho_{L-}\mathbf{T}_L(2)\rho_{L-} \tag{4.8}$$

has a similar result, the only change being that in 4.6 each ρ_{L+} becomes ρ_{L-}. The subalgebra $\rho_{L-}\mathbf{T}_L(2)\rho_{L-}$ is also isomorphic to $\mathbf{C}(16)$, and its object space is the antimatter subspace $\rho_-\mathbf{T}^2$ of $\mathbf{T}^2$. $\rho_{L\pm}\mathbf{T}_L(2)\rho_{L\pm}$ account for only half of $\mathbf{T}_L(2)$. The remaining elements mix the matter and antimatter subspaces of $\mathbf{T}^2$. More on this in chapter 6.

On specific elements of $\mathbf{R}_{1,9}$ the reductions 4.6 and 4.8 are illuminating in their distinctions. For example, the (1,9)-Dirac operator,

$$\boxed{\not{\partial}_{1,9} = \beta\partial^0 - \textstyle\sum_1^3 e_{L7}q_j\omega\partial^j - \sum_1^6 ie_{Lp}\omega\partial^{p+3},} \tag{4.9}$$

reduces via 4.6 to the (1,3)-matter-Dirac operator

$$\boxed{\not{\partial}_{1,3+} = \rho_{L+}\not{\partial}_{1,9}\rho_{L+} = (\beta\partial^0 + \textstyle\sum_1^3 iq_j\omega\partial^j)\rho_{L+} = (\sum_0^3 \gamma_\mu\partial^\mu)\rho_{L+}} \tag{4.10}$$

(the signs in 4.9 were arranged so that the signs in 4.10 would be conventional), and via 4.8 to the (1,3)-antimatter-Dirac operator

$$\not{\partial}_{1,3-} = \rho_{L-}\not{\partial}_{1,9}\rho_{L-} = (\beta\partial^0 - \sum_1^3 iq_j\omega\partial^j)\rho_{L-}. \tag{4.11}$$

Likewise, if

$$\tilde{\gamma}_5 = i(\beta)(-e_{L7}q_1\omega)(-e_{L7}q_2\omega)(-e_{L7}q_3\omega) = ie_{L7}\alpha,$$

then

$$\rho_{L\pm}\frac{1}{2}(\epsilon + \tilde{\gamma}_5)\rho_{L\pm} = \frac{1}{2}(\epsilon \pm \gamma_5)\rho_{L\pm}. \tag{4.12}$$

This associates left(right)handed matter with right(left)handed antimatter, and of course that is the observed association.

Perhaps a word of explication is needed here. The matrices

$$\gamma_0 = \beta,\ \gamma_j = iq_j\omega, j = 1,2,3,$$

via the maps

$$1 = q_0 \to \begin{bmatrix} 1 & 0 \\ 0 & 1 \end{bmatrix},\ iq_1 \to \begin{bmatrix} 0 & 1 \\ 1 & 0 \end{bmatrix},\ iq_2 \to \begin{bmatrix} 1 & 0 \\ 0 & -1 \end{bmatrix},\ iq_3 \to \begin{bmatrix} 0 & i \\ -i & 0 \end{bmatrix},$$

transform to

$$\gamma_0 \to \begin{bmatrix} . & . & 1 & 0 \\ . & . & 0 & 1 \\ 1 & 0 & . & . \\ 0 & 1 & . & . \end{bmatrix},\ \gamma_1 \to \begin{bmatrix} . & . & 0 & 1 \\ . & . & 1 & 0 \\ 0 & -1 & . & . \\ 1 & 0 & . & . \end{bmatrix},$$

$$\gamma_2 \to \begin{bmatrix} . & . & 1 & 0 \\ . & . & 0 & -1 \\ -1 & 0 & . & . \\ 0 & 1 & . & . \end{bmatrix},\ \gamma_3 \to \begin{bmatrix} . & . & 0 & i \\ . & . & -i & 0 \\ 0 & -i & . & . \\ i & 0 & . & . \end{bmatrix},$$

which are Dirac γ-matrices as commonly represented. The context with which we are dealing precludes the use of these conventional γ-matrices. We must play the hand we are dealt, and that is $\mathbf{T}_L(2) \simeq \mathbf{C} \otimes \mathbf{R}_{1,9}$. Let

$$\tilde{\gamma}_0 = \beta,\ \tilde{\gamma}_j = -e_{L7}q_j\omega, j = 1,2,3,\ \tilde{\gamma}_h = -ie_{Lh}\omega, h = 4,...,9,$$

be the $\mathbf{R}_{1,9}$ generators (1-vector basis) listed in 4.4 (signs specified to accomodate 4.9, leading to 4.10). The (1,3)-part of this basis consists of the $\tilde{\gamma}_\mu, \mu = 0,1,2,3,$ but at this point these operators are not matter/antimatter specific - the same set is valid for both. In making the projections

$$\tilde{\gamma}_0 \to \rho_{L\pm}\tilde{\gamma}_0\rho_{L\pm} = \gamma_0\rho_{L\pm},$$

$$\tilde{\gamma}_j \to \rho_{L\pm}\tilde{\gamma}_j\rho_{L\pm} = \pm\gamma_j\rho_{L\pm}, j = 1,2,3,$$

we achieve the more conventional γ_μ, with the antimatter γ's space reflected relative to the matter γ's. In addition, the matter/antimatter γ's are associated with the matter/antimatter projectors, $\rho_{L\pm}$. They do not appear without these projectors. And finally,

$$\tilde{\gamma}_h \to \rho_{L\pm}\tilde{\gamma}_h\rho_{L\pm} = 0,\ h = 4,...,9.$$

The matter/antimatter projectors kill the extra six dimensions. We shall look at these issues in greater detail in chapter 6.

4.2. Connecting to Particles.

Let Ψ be an element of $\mathbf{T}^2$ transforming with respect to $U(2)\times U(3)$ like $\Gamma_0^\dagger$ (any $\Gamma_m^\dagger$ would do). That is,

$$\begin{array}{llllll} \Gamma_0 & \longrightarrow_{U(1)} & e^{-\theta e_7/2}\Gamma_0 & \Rightarrow & \Psi \longrightarrow_{U(1)} & \Psi e^{\theta e_7/2}; \\ \Gamma_0 & \longrightarrow_{SU(2)} & e^{\tilde{\alpha}/2}\Gamma_0 & \Rightarrow & \Psi \longrightarrow_{SU(2)} & \Psi e^{-\tilde{\alpha}/2} \end{array} \tag{4.13}$$

(the factor 1/2 sets a conventional scale), so

$$\Psi\Gamma_0 \longrightarrow_{U(2)} \Psi\Gamma_0 \quad \text{(invariance)} \tag{4.14}$$

(recall that

$$\tilde{\alpha} = \alpha^1 ie_7q_1 + \alpha^2 q_2 + \alpha^3 ie_7 q_3 \Rightarrow \tilde{\alpha}\rho_\pm = (\pm\alpha^1 q_1 + \alpha^2 q_2 \pm \alpha^3 q_3)\rho_\pm,$$

hence it is natural to associate the $U(1)$ generator $(-e_7)$ with $\tilde{\alpha}$, since $-e_7\rho_\pm = \pm i\rho_\pm$). It will be understood henceforth that by the symbol Ψ we mean the canonical $\mathbf{T}^2$ spinor such that $\Psi\Gamma_0 = \Psi\Delta_0$ (modulo $U(3)$). However, even though this will be understood, the former notation will be used to avoid confusion with the Δ_0 in the decomposition $\Psi = \sum_0^3 \Psi\Delta_m$.

With respect to $U(3)$ the situation is different, for all the Γ_m are $U(3)$ invariant. The action of $U(1) \subset U(3)$ on Ψ is the inverse of 3.64:

$$\begin{aligned} \Psi & \rightarrow e^{\phi e_7/6}\Psi e^{-\phi e_7/6} \\ & = \rho_+\Psi\rho_+ + \rho_-\Psi\rho_- + e^{-i\phi/3}\rho_+\Psi\rho_- + e^{i\phi/3}\rho_-\Psi\rho_+. \end{aligned} \tag{4.15}$$

With respect to $SU(3)$ Ψ transforms as $\mathbf{1}\oplus\bar{\mathbf{1}}\oplus\mathbf{3}\oplus\bar{\mathbf{3}}$ (corresponding to the respective terms in 4.15). In particular,

$$\rho_+\Psi = \rho_+[\Psi_\mathbf{1} + \Psi_\mathbf{3}] = \rho_+[\Psi_\mathbf{1} + \Psi_\mathbf{r}e_1 + \Psi_\mathbf{g}e_2 + \Psi_\mathbf{b}e_4], \tag{4.16}$$

where $\Psi_\mathbf{1}$ and the Ψ_a, $a = \mathbf{r},\mathbf{g},\mathbf{b}$, each consist of an $SU(2)$ doublet of Dirac spinors. The subscripts $\mathbf{r}$= red, $\mathbf{g}$= green, $\mathbf{b}$= blue, are conventional $SU(3)$ 'color' labels. We assign them conventional $SU(3)$ color charges:

$$\mathbf{r} = (\tfrac{1}{2},\tfrac{1}{3}), \quad \mathbf{g} = (-\tfrac{1}{2},\tfrac{1}{3}), \quad \mathbf{b} = (0,-\tfrac{2}{3}).$$

Let I_3 be the quantum number of $SU(2)$. It is associated with the action

$$I_3\Psi\lambda_\pm = \Psi\lambda_\pm i\tilde{x}/2 = \pm\frac{1}{2}\Psi\lambda_\pm.$$

Y_2, the hypercharge of $U(1) \subset U(2)$, is associated with the action,

$$Y_2\Psi\rho_\pm = \Psi\rho_\pm(-ie_7/2) = \mp\frac{1}{2}\Psi\rho_\pm,$$

and Y_3, the hypercharge of $U(1) \subset U(3)$, with the action

$$Y_3\rho_+\Psi\rho_\pm = (-ie_7/6)\rho_+\Psi\rho_\pm + \rho_+\Psi\rho_\pm(ie_7/6) = 0 \text{ or } -\frac{1}{3}\rho_+\Psi\rho_-.$$

Below is a table of values for I_3, Y_2, Y_3 for the $SU(3)$ singlet and triplet parts of Ψ, together with the charge $Q = I_3 + Y_2 + Y_3$.

$$\begin{array}{lcccc} & I_3 & Y_2 & Y_3 & Q \\ \rho_+\Psi\rho_+\lambda_+ = \Psi_{\mathbf{1}}\Delta_0 & +\frac{1}{2} & -\frac{1}{2} & 0 & 0 \\ \rho_+\Psi\rho_+\lambda_- = \Psi_{\mathbf{1}}\Delta_1 & -\frac{1}{2} & -\frac{1}{2} & 0 & -1 \\ \rho_+\Psi\rho_-\lambda_+ = \Psi_{\mathbf{3}}\Delta_2 & +\frac{1}{2} & +\frac{1}{2} & -\frac{1}{3} & +\frac{2}{3} \\ \rho_+\Psi\rho_-\lambda_- = \Psi_{\mathbf{3}}\Delta_3 & -\frac{1}{2} & +\frac{1}{2} & -\frac{1}{3} & -\frac{1}{3} \end{array} \tag{4.17}$$

The obvious particle assignments are:

$$\begin{aligned} \rho_+\Psi\rho_+\lambda_+ &\rightarrow \text{neutrino}, \\ \rho_+\Psi\rho_+\lambda_- &\rightarrow \text{electron}, \\ \rho_+\Psi\rho_-\lambda_+ &\rightarrow \text{up-quark}, \\ \rho_+\Psi\rho_-\lambda_- &\rightarrow \text{down-quark}. \end{aligned} \tag{4.18}$$

The antifamily charges are

$$\begin{array}{lcccc} & I_3 & Y_2 & Y_3 & Q \\ \rho_-\Psi\rho_-\lambda_- = \Psi_{\bar{\mathbf{1}}}\Delta_3 & -\frac{1}{2} & +\frac{1}{2} & 0 & 0 \\ \rho_-\Psi\rho_-\lambda_+ = \Psi_{\bar{\mathbf{1}}}\Delta_2 & +\frac{1}{2} & +\frac{1}{2} & 0 & +1 \\ \rho_-\Psi\rho_+\lambda_- = \Psi_{\bar{\mathbf{3}}}\Delta_1 & -\frac{1}{2} & -\frac{1}{2} & +\frac{1}{3} & -\frac{2}{3} \\ \rho_-\Psi\rho_+\lambda_+ = \Psi_{\bar{\mathbf{3}}}\Delta_0 & +\frac{1}{2} & -\frac{1}{2} & +\frac{1}{3} & +\frac{1}{3} \end{array} \tag{4.19}$$

In fact, however, due to parity nonconservation the quantum numbers 4.17, and the particle assignments 4.18, are not entirely correct. Only the lefthanded Weyl spinor parts of the associated Dirac spinors have these quantum numbers. The righthanded spinors are $SU(2)$ singlets ($I_3 = 0$), and have different values of Y_2.

Spinors with the correct righthanded quantum numbers arise as follows. The inner product of Ψ with any other spinor in $\mathbf{T}^2$, a la equation 3.39, associates Ψ with the Γ_m. New spinors, $\Psi\Gamma_m$, arise in this way. The action of $SU(2)$ on Ψ was defined so as to cancel the action on Γ_m:

$$\Psi\Gamma_m \to (\Psi e^{-\tilde{\alpha}/2})(e^{\tilde{\alpha}/2}\Gamma_m) = \Psi\Gamma_m. \tag{4.20}$$

So each of the eight spinors $\rho_\pm\Psi\Gamma_m$ is $SU(2)$ chargeless.

The Y_2 charges for the Γ_m are carried by the automorphisms M_m from Γ_0 to Γ_j, j=1,2,3 (see 3.62). Therefore, if

$$Y_2(\Gamma_0) = +1/2,$$

then

$$Y_2(\Gamma_1) = -1/2,$$

$$Y_2(\Gamma_2) = +1/2,$$

$$Y_2(\Gamma_3) = -1/2.$$

The $U(3)$ charges of $\Psi\Gamma_m$ are the same as those for $\Psi\Delta_m$ above. $SU(3)$ is a subgroup of the automorphism group of $\mathbf{O}$, so its action on $\rho_\pm\Psi\Gamma_m$ is the same as that on $\rho_\pm\Psi\Delta_m$, as the only term in either not $SU(3)$ invariant is Ψ. Likewise, the Y_3 charges are the same since e_7 commutes with Γ_m, and if $e_7\Delta_m = \pm i\Delta_m$, then $e_7\Gamma_m = \pm i\Gamma_m$ too.

In particular, it should be emphasized that the fact that parity nonconservation occurs with respect to $SU(2)$, but not $SU(3)$, is being demonstrated to arise from the fact that the Γ_m are not $SU(2)$ invariant, but are $SU(3)$ invariant.

Quantum numbers for the $\rho_+\Psi\Gamma_m$ are listed below:

$$\begin{array}{lcccc} & I_3 & Y_2 & Y_3 & Q \\ \rho_+\Psi\Gamma_0 = \Psi_{\mathbf{1}}\Gamma_0 & 0 & 0 & 0 & 0 \\ \rho_+\Psi\Gamma_1 = \Psi_{\mathbf{1}}\Gamma_1 & 0 & -1 & 0 & -1 \\ \rho_+\Psi\Gamma_2 = \Psi_{\mathbf{3}}\Gamma_2 & 0 & +1 & -\frac{1}{3} & +\frac{2}{3} \\ \rho_+\Psi\Gamma_3 = \Psi_{\mathbf{3}}\Gamma_3 & 0 & 0 & -\frac{1}{3} & -\frac{1}{3} \end{array} \tag{4.21}$$

For example, under the action of the $U(1)$ of Y_2,

$$\begin{aligned} \rho_+\Psi\Gamma_2 &\rightarrow \rho_+\Psi e^{\theta e_7/2}e^{i\theta/2}\Gamma_2 \\ &= \rho_+\Psi\Gamma_2\rho_- e^{\theta(e_7+i)/2} \quad (\Gamma_2 = \Gamma_2\rho_-) \\ &= \rho_+\Psi\Gamma_2\rho_- e^{\theta(i+i)/2} \quad (\rho_- e_7 = i\rho_-) \\ &= \rho_+\Psi\Gamma_2 e^{i\theta}. \end{aligned}$$

The quantum numbers 4.21 give rise to the same particle assignments as 4.17, but this time they are those of righthanded leptons and quarks. The question therefore arises, how can we mathematically associate the quantum numbers 4.17 with

$$\boxed{\tfrac{1}{2}(\epsilon + \gamma_5)\rho_+\Psi\Delta_m = \text{ lefthanded family,}} \tag{4.22}$$

and the quantum numbers 4.21 with

$$\boxed{\tfrac{1}{2}(\epsilon - \gamma_5)\rho_+\Psi\Gamma_m = \text{ righthanded family?}} \tag{4.23}$$

The antifamily charges corresponding to 4.21 are

$$\begin{array}{lcccc} & I_3 & Y_2 & Y_3 & Q \\ \rho_-\Psi\Gamma_3 = \Psi_{\bar{\mathbf{1}}}\Gamma_3 & 0 & 0 & 0 & 0 \\ \rho_-\Psi\Gamma_2 = \Psi_{\bar{\mathbf{1}}}\Gamma_2 & 0 & +1 & 0 & +1 \\ \rho_-\Psi\Gamma_1 = \Psi_{\bar{\mathbf{3}}}\Gamma_1 & 0 & -1 & +\frac{1}{3} & -\frac{2}{3} \\ \rho_-\Psi\Gamma_0 = \Psi_{\bar{\mathbf{3}}}\Gamma_0 & 0 & 0 & +\frac{1}{3} & +\frac{1}{3} \end{array} \tag{4.24}$$

In this case we'd want to associate the charges 4.19 to

$$\boxed{\tfrac{1}{2}(\epsilon - \gamma_5)\rho_-\Psi\Delta_m = \text{ righthanded antifamily,}} \tag{4.25}$$

and those in 4.24 to

$$\boxed{\tfrac{1}{2}(\epsilon + \gamma_5)\rho_-\Psi\Gamma_m = \text{ lefthanded antifamily.}} \tag{4.26}$$

4.3. Parity Nonconservation.

RIGHTHANDED DIRAC OPERATOR

Define

$$\begin{aligned} L_\pm &= \frac{1}{2}(\epsilon + \gamma_5)\rho_{L\pm}, \\ R_\pm &= \frac{1}{2}(\epsilon - \gamma_5)\rho_{L\pm}. \end{aligned} \tag{4.27}$$

Define

$$\boxed{\partial\!\!\!/_{R+} = \partial\!\!\!/_{1,3}R_+ = \begin{bmatrix} 0 & \sum_0^3 \sigma_\mu\partial^\mu \\ 0 & 0 \end{bmatrix} \rho_{L+},} \tag{4.28}$$

where $\sigma_0 = 1$, and $\sigma_j = iq_j$, j=1,2,3. Let Ψ in $\mathbf{T}^2$ be a space-time field (for now; later we'll consider Ψ a (1,9)-space-time field). Consider the inner product

$$< \Psi, \partial\!\!\!/_{R+}\Psi >= \frac{1}{8}\{[\sum_{mn} M_m((\Psi\Gamma_n)^\wedge(\partial\!\!\!/_{R+}\Psi\Gamma_n))] + [same]^\dagger\} \tag{4.29}$$

where

$$\Psi = \begin{bmatrix} \Psi_L \\ \Psi_R \end{bmatrix}, \tag{4.30}$$

and

$$\Psi^\wedge = \begin{bmatrix} \Psi_L \\ \Psi_R \end{bmatrix}^\wedge = [\ \Psi_R^\dagger \quad \Psi_L^\dagger\ ], \tag{4.31}$$

is the spinor conjugate. In 4.29 consider the term

$$\partial\!\!\!/_{R+}\Psi\Gamma_n = \begin{bmatrix} 0 & \sum_0^3 \rho_+\sigma_\mu\partial^\mu\Psi_R\Gamma_n \\ 0 & 0 \end{bmatrix}. \tag{4.32}$$

The derivative in 4.32 'sees' both Ψ_R and Γ_n, so it sees the $U(2) \times U(3)$ actions on each. That is, in making this internal symmetry local, gauge fields arise consistent with the charges 4.21 associated with righthanded leptoquark fields. Note that $SU(3)$ gauge fields survive because the Γ_n are invariant, and so have no effect on the gauge fields arising from the action of $SU(3)$ on Ψ.

LEFTHANDED DIRAC OPERATOR

A trick is used to manifest the correct gauge fields for lefthanded leptons and quarks. Define

$$\boxed{\partial\!\!\!/_{L+} = \begin{bmatrix} 0 & 0 \\ \sum_0^3 \sigma_\mu^- \overleftarrow{\partial^\mu} & 0 \end{bmatrix} \rho_{L+},} \tag{4.33}$$

where $\sigma_0^- = \sigma_0$, $\sigma_j^- = -\sigma_j$, j=1,2,3, and

$$\overleftarrow{\partial^\mu}(A) = A\overleftarrow{\partial^\mu}$$

is differentiation from the right. That is, it is assumed that since

$$\begin{bmatrix} 0 & 0 \\ \sigma_\mu^- & 0 \end{bmatrix} \rho_{L+} = M_1(\begin{bmatrix} 0 & \sigma_\mu \\ 0 & 0 \end{bmatrix} \rho_{L+})^\dagger, \tag{4.34}$$

where $A \rightarrow M_1(A)^\dagger$ is an antiautomorphism (product reversing), that this reverses the direction of the action of the derivative (ie., under the action of this antiautomorphism,

$$\partial^\mu A \rightarrow M_1[\partial^\mu M_1(A)^\dagger]^\dagger = A\overleftarrow{\partial^\mu}).$$

Take the inner product

$$< \Psi, \partial\!\!\!/_{L+}\Psi > = \frac{1}{8}\{[\sum_{mn} M_m((\Psi\Gamma_n)^\wedge(\partial\!\!\!/_{L+}\Psi\Gamma_n))] + [same]^\dagger\}, \tag{4.35}$$

and consider the term

$$\partial\!\!\!/_{L+}\Psi\Gamma_n = \begin{bmatrix} 0 & 0 \\ \sum_0^3 \rho_+\sigma_\mu^- \Psi_L \overleftarrow{\partial^\mu}\Gamma_n & 0 \end{bmatrix}. \tag{4.36}$$

In 4.36 the derivative $\overleftarrow{\partial^\mu}$ sees only

$$\Psi_L = \sum_0^3 \Psi_L \Delta_m,$$

hence only the $U(2) \times U(3)$ actions on this field. Therefore gauge fields associated with the charges 4.17 arise in this case, those of lefthanded leptons and quarks. $SU(3)$ gauge fields are the same as the righthanded case.

FULL PARITY VIOLATING DIRAC OPERATOR

The operator

$$\boxed{\partial\!\!\!/_+ = \partial\!\!\!/_{L+} + \partial\!\!\!/_{R+}} \tag{4.37}$$

differs from $\partial\!\!\!/_{1,3+}$ (our earlier (1,3)-matter-Dirac operator defined in 4.10) only in the direction of the derivative in the lefthanded part, so it is also a valid (1,3)-matter-Dirac operator. Its advantage, of course, is that its use in the spinor Lagrangian term

$$\boxed{\mathcal{L}_0 = <\Psi, \partial\!\!\!/_+\Psi> = \tfrac{1}{8}\{[\textstyle\sum_{mn} M_m((\Psi\Gamma_n)^\wedge(\partial\!\!\!/_+\Psi\Gamma_n))] + [same]^\dagger\}} \tag{4.38}$$

results after gauging in the observed lepto-quark gauge field interactions. To this point the only unconventional result is the appearance of a chargeless righthanded neutrino. And, of course, the structures being manifested are intrinsic to the mathematics.

Because of the appearance of ρ_+ in $\partial\!\!\!/_+$, $\mathcal{L}_0$ is only a matter Lagrangian, and it gives rise to only matter to matter transitions. That is,

$$\mathcal{L}_0 = <\Psi, \partial\!\!\!/_+\Psi> = <\rho_+\Psi, \partial\!\!\!/_+\rho_+\Psi> .$$

The $\rho_-\Psi$ parts of Ψ fall out in taking the inner product, because in general

$$<\rho_\pm A, B> = <A, \rho_\pm B> = <\rho_\pm A, \rho_\pm B>$$

($\rho_\pm$ could be replaced by any Hermitian associative projection operator). An antimatter version of $\mathcal{L}_0$ is easily constructed along the lines outlined above, but in the next few sections attention will be restricted to matter and $\mathcal{L}_0$.

4.4. Gauge Fields.

The core of the Lagrangian $\mathcal{L}_0$ consists of the terms

$$(\Psi\Gamma_n)^\wedge(\not{\partial}_+\Psi\Gamma_n) = (\Psi_R\Gamma_n)^\dagger(\partial\rho_+\Psi_R\Gamma_n) + (\Psi_L\Gamma_n)^\dagger(\partial^-\rho_+\Psi_L\Gamma_n), \quad (4.39)$$

where

$$\partial = \sum \sigma_\mu\partial^\mu, \ \partial^- = \sum \sigma_\mu^-\overleftarrow{\partial^\mu}.$$

Define

$$\boxed{\tau_1 = q_1, \ \tau_2 = q_1\tilde{x}, \ \tau_3 = \tilde{x}.} \quad (4.40)$$

These are alternate $SU(2)$ generators, and $\tau_3 = \tilde{x}$ is diagonal with respect to the projection operators $\lambda_\pm = \frac{1}{2}(1 \pm i\tilde{x})$. Consider the n=0 term of the lefthanded part of 4.39. $U(2)$ gauge fields arise as follows ($\rho_+\Psi\Gamma_0$ is $U(3)$ invariant):

$$\begin{aligned}
&(\Psi_L\Gamma_0)^\dagger(\sigma_\mu^-\rho_+\Psi_L[\tfrac{1}{2}gA^{\mu j}\tau_j - \tfrac{1}{2}g'B^\mu e_7]\Gamma_0) \\
= \ &(\Psi_L\Gamma_0)^\dagger(\sigma_\mu^-\rho_+\Psi_L(\Delta_0+\Delta_1)[\tfrac{1}{2}gA^{\mu j}\tau_j + \tfrac{1}{2}g'iB^\mu]\Gamma_0) \quad (\Delta_2, \Delta_3 \text{ terms drop ou} \\
= \ &(\Psi_L\Gamma_0)^\dagger(\sigma_\mu^-\rho_+\Psi_L\Delta_0[\tfrac{1}{2}gA^{\mu 3}\tilde{x} + \tfrac{1}{2}g'iB^\mu]\Gamma_0) \quad (\Delta_0(\tau_1+\tau_2)\Gamma_0 = 0) \\
&+(\Psi_L\Gamma_0)^\dagger(\sigma_\mu^-\rho_+\Psi_L\Delta_1[\tfrac{1}{2}g(A^{\mu 1}q_1 + A^{\mu 2}q_1\tilde{x})]\Gamma_0) \quad (\Delta_1(\tau_3+i)\Gamma_0 = 0) \\
\approx \ &(N_L\Gamma_0)^\dagger(\sigma_\mu^- N_L[\tfrac{1}{2}\Delta_0 i(g'B^\mu - gA^{\mu 3})]\Gamma_0) \quad (\Delta_0\tilde{x}\Gamma_0 = -i\Gamma_0) \\
&+(N_L\Gamma_0)^\dagger(\sigma_\mu^- E_L[\tfrac{1}{2}\Delta_1 q_1 g(A^{\mu 1} - iA^{\mu 2})]\Gamma_0)
\end{aligned} \quad (4.41)$$

(summation over repeated indices), where

$$N_L = \rho_+\Psi_L\Delta_0$$

is a lefthanded neutrino field, and

$$E_L = \rho_+\Psi_L\Delta_1$$

is a lefthanded charged lepton field. The last step in 4.41 is not a strict equality because $(\Psi_L\Gamma_0)^\dagger = ((\rho_+ + \rho_-)\Psi_L\Gamma_0)^\dagger$, but in completing the inner product 4.38 the $(\rho_-\Psi_L\Gamma_0)^\dagger$ term drops out. g and g' are conventional gauge coupling constants for $SU(2)$ and $U(1)$.

In like manner, setting n=1 in the lefthanded part of 4.39, we obtain

$$(\Psi_L\Gamma_1)^\dagger(\sigma_\mu^- \rho_+ \Psi_L[\tfrac{1}{2}gA^{\mu j}\tau_j - \tfrac{1}{2}g'B^\mu e_7]\Gamma_1)$$

$$\approx (E_L\Gamma_1)^\dagger(\sigma_\mu^- E_L[\tfrac{1}{2}\Delta_1 i(g'B^\mu + gA^{\mu 3})]\Gamma_1) \tag{4.42}$$

$$+(E_L\Gamma_1)^\dagger(\sigma_\mu^- N_L[\tfrac{1}{2}\Delta_0 q_1 g(A^{\mu 1} + iA^{\mu 2})]\Gamma_1).$$

Another word in explication is needed here. The last lines of 4.41 are those parts of the first line that survive the inner product. The gauge field terms, and the indecomposable (with respect to $U(2) \times U(3)$) Pauli spinor bits of Ψ, are all easily identifiable with physical particles. Combinations of these algebraic pieces of the gauged Lagrangian that survive the inner product correspond to viable transitions; those that do not are not viable. *They will not occur.* No actual cross-sections or probabilities are being calculated here. As yet we may only read from the mathematics a binary result: yes, this transition is allowed; no, that one is not.

The algebraic gauge fields are indicated by square brackets in the last terms of 4.41 and 4.42, and from these we can figure their Y_2, I_3 (and $Q = Y_2 + I_3$) charges ($U(3)$ invariant). Only two of these charges are nonzero. The first is:

$$I_3[\frac{1}{2}\Delta_1 q_1(A^{\mu 1} - iA^{\mu 2})] = -\frac{1}{2}i\tilde{x}[...] + [...]\frac{1}{2}i\tilde{x} = \frac{1}{2}[...] + [...]\frac{1}{2} = [...], \tag{4.43}$$

where we have exploited the fact that

$$-\frac{1}{2}i\tilde{x}\Delta_1 q_1 + \Delta_1 q_1 \frac{1}{2}i\tilde{x} = \Delta_1 q_1.$$

So $I_3 = 1$ for this gauge field, and

$$Q = I_3 + Y_2 = 1 + 0 = 1,$$

and we make the conventional identification

$$W^{+\mu} = (A^{\mu 1} - iA^{\mu 2})/2^{\frac{1}{2}}. \tag{4.44}$$

So the last term in 4.41 represents the transition

$$E_L + W^+ \rightarrow N_L. \tag{4.45}$$

Likewise,

$$W^{-\mu} = (A^{\mu 1} + iA^{\mu 2})/2^{\frac{1}{2}}, \tag{4.46}$$

$(I_3 = -1, Y_2 = 0)$ and the last term in 4.42 corresponds to the transition

$$N_L + W^- \rightarrow E_L. \tag{4.47}$$

It will be noted that the issue of second quantization has not yet arisen. When algebraic fields identifiable with particular particles (by their symmetry transformation properties) appear in combination in a Lagrangian, and the combination survives the inner product, then an allowable transition results. Part of the $U(2) \times U(3)$ gauging of $\mathcal{L}_0$, denote it $\mathcal{L}_g$ (ie., $\mathcal{L}_g$ is distinct from $\mathcal{L}_0$ and contains all the gauge field-fermion interaction terms), is

$$< E_L, \sigma_\mu^- N_L[\frac{1}{\sqrt{2}}\Delta_0 q_1 g W^{\mu -}] > \neq 0,$$

which in having a nonzero real part implies 4.47 is an allowable transition. On the other hand,

$$< E_L, \sigma_\mu^- E_L[\frac{1}{\sqrt{2}}\Delta_0 q_1 g W^{\mu -}] > = 0,$$

so $E_L \rightarrow E_L + W^-$ is not allowed.

Two other conventional definitions are made:

$$Z^\mu = \frac{g' B^\mu - g A^{\mu 3}}{(g^2 + g'^2)^{1/2}} \tag{4.48}$$

and

$$A^\mu = \frac{g B^\mu + g' A^{\mu 3}}{(g^2 + g'^2)^{1/2}}, \tag{4.49}$$

so

$$g' B^\mu + g A^{\mu 3} = 2\frac{g g'}{(g^2 + g'^2)^{1/2}} A^\mu + \frac{g'^2 - g^2}{(g^2 + g'^2)^{1/2}} Z^\mu \tag{4.50}$$

(with

$$e = \frac{g g'}{(g^2 + g'^2)^{1/2}}$$

the electromagnetic coupling constant). These arise in the terms

$$(N_L \Gamma_0)^\dagger (\sigma_\mu^- N_L[\frac{1}{2}\Delta_0 q_1 (g^2 + g'^2)^{1/2} Z^\mu] \Gamma_0) \tag{4.51}$$

from 4.41 (neutral current), and

$$(E_L\Gamma_1)^\dagger(\sigma_\mu^- E_L[\Delta_1 i(eA^\mu + \frac{g'^2 - g^2}{2(g^2+g'^2)^{1/2}} Z^\mu]\Gamma_1) \tag{4.52}$$

from 4.42 (electromagnetic and neutral currents).

The righthanded term in 4.39 also gauges in the conventional manner. However, this term contains the following:

$$(\rho_+\Psi_R\Gamma_0)^\dagger(\partial\rho_+\Psi_R\Gamma_0) = N_R^\dagger(\partial N_R) \tag{4.53}$$

($N_R = \rho_+\Psi_R\Gamma_0$). N_R is a righthanded neutrino. Since it is $U(2)\times U(3)$ chargeless, it does not appear in $\mathcal{L}_g$. The presence of a righthanded neutrino in this theory is a prediction, along with all that it would imply in astrophysics, in particular regarding the problem of dark matter. Its only interaction may be gravitational.

4.5. Weak Mixing.

The symmetry of equation 3.40 is larger than $U(2)\times U(3)$. There are extra $U(1)$'s and an extra $SU(2)$ (see 3.47). Let

$$\Gamma_0 = U_1U_2\Delta_0, \tag{4.54}$$

$U_1 \in U(1)$, $U_2 \in SU(2)$. The remaining $\Gamma_j, j = 1,2,3$, differ from $M_j(\Gamma_0)$ by the extra $U(1)$ and $SU(2)$ phases. Let

$$\tilde{\beta} = \beta^1 ie_7q_1 + \beta^2 q_2 + \beta^3 ie_7q_3,$$

and let

$$P = exp(\tilde{\beta}\rho_-).$$

Therefore,

$$P\Gamma_m = \Gamma_m,\ m = 0,1,$$

$$P\Gamma_m = e^{\tilde{\beta}}\Gamma_m,\ m = 2,3$$

(see 3.48). Since P has a nontrivial effect only on Γ_2 and Γ_3, it can be used to account for the additional $SU(2)$ freedom in the Γ_m. We may express the

Γ_m as follows:

$$\begin{aligned}\Gamma_0 &= U_1U_2\Delta_0 = U_1(PU_2P^{-1})[P\Delta_0],\\ \Gamma_1 &= e^{i\theta_1}M_1(\Gamma_0) = U_1(PU_2P^{-1})[e^{i\theta_1}P\Delta_1],\\ \Gamma_2 &= e^{i\theta_2}M_2(\Gamma_0) = U_1(PU_2P^{-1})[e^{i\theta_2}P\Delta_2],\\ \Gamma_3 &= e^{i\theta_3}M_3(\Gamma_0) = U_1(PU_2P^{-1})[e^{i\theta_3}P\Delta_3].\end{aligned} \tag{4.55}$$

In the square brackets are incorporated the additional symmetry bits, but outside there appears a revamped

$$SU(2) \rightarrow PU_2P^{-1}.$$

Since

$$PU_2P^{-1}\Gamma_m = U_2\Gamma_m,\ m = 0, 1,$$

the $SU(2)$ gauge fields of the previous section are unaltered by this change. This is true even if we alter the $U(2)$ action on the matter spinor field $\rho_+\Psi$:

$$\begin{aligned}\rho_+\Psi \quad &\rightarrow_{U(2)} \quad \rho_+\Psi(PU_2^{-1}P^{-1})U_1^{-1}\\ &= \quad \rho_+\Psi\Delta_0U_2^{-1}U_1^{-1} + \rho_+\Psi\Delta_1U_2^{-1}U_1^{-1}\\ &\quad +\rho_+\Psi\Delta_2(e^{\tilde{\beta}}U_2^{-1}e^{-\tilde{\beta}})U_1^{-1} + \rho_+\Psi\Delta_3(e^{\tilde{\beta}}U_2^{-1}e^{-\tilde{\beta}})U_1^{-1}.\end{aligned} \tag{4.56}$$

That is, the neutral and charged lepton fields, $\rho_+\Psi\Delta_0$ and $\rho_+\Psi\Delta_1$, are unaltered. Not so the quark fields, $\rho_+\Psi\Delta_2$ and $\rho_+\Psi\Delta_3$.

It is necessary in what follows to specify that the $\theta_j, j = 1, 2, 3$, and the components of $\tilde{\beta}$, are to be treated as constant phases. Only the $U(2) \times U(3)$ previously defined will give rise to gauge fields. For the moment the justification for this assumption will rest on its consequences.

Making use of 4.55,56, the lefthanded n=2 part of the $\mathcal{L}_0$ terms 4.39 transform into

$$\begin{aligned}&\{(\Psi_L(PU_2^{-1}P^{-1})U_1^{-1}U_1(PU_2P^{-1})[e^{i\theta_2}P\Delta_2])^\dagger\times\\ &(\sigma_\mu^-\rho_+\Psi_L(PU_2^{-1}P^{-1})U_1^{-1}\overleftarrow{\partial^\mu}U_1(PU_2P^{-1})[e^{i\theta_2}P\Delta_2])\}\\ &\approx \{(\rho_+\Psi_L[P\Delta_2])^\dagger(\sigma_\mu^-\rho_+\Psi_LPU_2^{-1}U_1^{-1}\overleftarrow{\partial^\mu}U_1U_2\Delta_2)\}.\end{aligned} \tag{4.57}$$

Including $U(1) \subset U(3)$, but ignoring $SU(3)$, this gauges as follows:

$$\begin{aligned}&(\rho_+\Psi_L P\Gamma_2)^\dagger(\sigma_\mu^- \rho_+\Psi_L P[\tfrac{1}{2}gA^{\mu j}\tau_j - \tfrac{1}{6}g'B^\mu e_7]\Gamma_2)\\ \approx\ &(\rho_+\Psi_L P\Gamma_2)^\dagger(\sigma_\mu^- \rho_+\Psi_L(\Delta_2+\Delta_3)P[\tfrac{1}{2}gA^{\mu j}\tau_j - \tfrac{1}{6}g'B^\mu e_7]\Gamma_2)\end{aligned} \tag{4.58}$$

(reverting back to the symbol Γ_2 instead of the Δ_2 appearing in 4.57). Now set

$$P = e^{\beta\tilde{x}/2} = e^{\beta\tau_3/2},$$

β real, so that

$$\Delta_2 P = \Delta_2 e^{-i\beta/2},$$

$$\Delta_3 P = \Delta_3 e^{+i\beta/2}.$$

Also,

$$P\Gamma_2 = e^{-i\beta/2}\Gamma_2$$

for the canonical Γ_2 $(= \Delta_2)$. In this case 4.58 reduces to

$$\begin{aligned}&(\rho_+\Psi_L e^{-i\beta/2}\Gamma_2)^\dagger(\sigma_\mu^- \rho_+\Psi_L\Delta_2 e^{-i\beta/2}[\tfrac{1}{2}gA^{\mu j}\tau_j - \tfrac{1}{6}g'B^\mu e_7]\Gamma_2)\\ &+(\rho_+\Psi_L e^{-i\beta/2}\Gamma_2)^\dagger(\sigma_\mu^- \rho_+\Psi_L\Delta_3 e^{+i\beta/2}[\tfrac{1}{2}gA^{\mu j}\tau_j - \tfrac{1}{6}g'B^\mu e_7]\Gamma_2)\\ &= (\rho_+\Psi_L\Gamma_2)^\dagger(\sigma_\mu^- \rho_+\Psi_L\Delta_2[i(-\tfrac{2}{3}eA^\mu + \tfrac{3g^2-g'^2}{6(g^2+g'^2)^{\frac{1}{2}}}Z^\mu]\Gamma_2)\\ &+(\rho_+\Psi_L\Gamma_2)^\dagger(\sigma_\mu^- \rho_+\Psi_L\Delta_3 e^{+i\beta}[\tfrac{1}{\sqrt{2}}q_1 gW^{+\mu}]\Gamma_2).\end{aligned} \tag{4.59}$$

$\rho_+\Psi\Delta_2$ is a $Q = \frac{2}{3}$ quark, $\rho_+\Psi\Delta_3$ a $Q = -\frac{1}{3}$ quark. Let

$$\rho_+\Psi\Delta_2 = U + iC,$$

U = up-quark, C = charm quark, and

$$\rho_+\Psi\Delta_3 = D + iS,$$

D = down-quark, S = strange quark. Then 4.59 can be written

$$
\begin{aligned}
&((U_L + iC_L)\Gamma_2)^{\dagger}(\sigma_{\mu}^{-}(U_L + iC_L)[i(-\tfrac{2}{3}eA^{\mu} + \tfrac{3g^2 - g'^2}{6(g^2+g'^2)^{\frac{1}{2}}} Z^{\mu}]\Gamma_2) \\
&+((U_L + iC_L)\Gamma_2)^{\dagger}(\sigma_{\mu}^{-}(D_L + iS_L)e^{i\beta}[\tfrac{1}{\sqrt{2}}q_1 g W^{+\mu}]\Gamma_2) \\
= \;& ((U_L)\Gamma_2)^{\dagger}(\sigma_{\mu}^{-}(U_L)[i(-\tfrac{2}{3}eA^{\mu} + \tfrac{3g^2 - g'^2}{6(g^2+g'^2)^{\frac{1}{2}}} Z^{\mu}]\Gamma_2) \\
&+(C_L\Gamma_2)^{\dagger}(\sigma_{\mu}^{-}(C_L)[i(-\tfrac{2}{3}eA^{\mu} + \tfrac{3g^2 - g'^2}{6(g^2+g'^2)^{\frac{1}{2}}} Z^{\mu}]\Gamma_2) \\
&+i\{(U_L\Gamma_2)^{\dagger}(\sigma_{\mu}^{-}(C_L)[i(-\tfrac{2}{3}eA^{\mu} + \tfrac{3g^2 - g'^2}{6(g^2+g'^2)^{\frac{1}{2}}} Z^{\mu}]\Gamma_2) \\
&-(C_L\Gamma_2)^{\dagger}(\sigma_{\mu}^{-}(U_L)[i(-\tfrac{2}{3}eA^{\mu} + \tfrac{3g^2 - g'^2}{6(g^2+g'^2)^{\frac{1}{2}}} Z^{\mu}]\Gamma_2)\} \\
&+(U_L\Gamma_2)^{\dagger}(\sigma_{\mu}^{-}(D_L cos(\beta) - S_L sin(\beta))[\tfrac{1}{\sqrt{2}}q_1 g W^{+\mu}]\Gamma_2) \\
&+(C_L\Gamma_2)^{\dagger}(\sigma_{\mu}^{-}(S_L cos(\beta) + D_L sin(\beta))[\tfrac{1}{\sqrt{2}}q_1 g W^{+\mu}]\Gamma_2) \\
&+i\{(U_L\Gamma_2)^{\dagger}(\sigma_{\mu}^{-}(S_L cos(\beta) + D_L sin(\beta))[\tfrac{1}{\sqrt{2}}q_1 g W^{+\mu}]\Gamma_2) \\
&-(C_L\Gamma_2)^{\dagger}(\sigma_{\mu}^{-}(D_L cos(\beta) - S_L sin(\beta))[\tfrac{1}{\sqrt{2}}q_1 g W^{+\mu}]\Gamma_2)\}.
\end{aligned}
\tag{4.60}
$$

By replacing

$$\rho_+\Psi = (\text{electron family}) + i(\text{muon family})$$

(so that $\rho_+\Psi\Delta_2 = U + iC$, and $\rho_+\Psi\Delta_3 = D + iS$) by

$$\rho_+\Psi' = (\text{electron family}) - i(\text{muon family}),$$

the terms in 4.60 delimited $i\{...\}$ change sign, so in adding the Ψ and Ψ' versions together, these terms drop out. What remains are the U-to-U and C-to-C electromagnetic and neutral currents, plus charged currents with families mixed. For example,

$$(U_L\Gamma_2)^\dagger(\sigma_\mu^-(D_L cos(\beta) - S_L sin(\beta))[\tfrac{1}{\sqrt{2}}q_1 g W^{+\mu}]\Gamma_2)$$

$$= (U_L\Gamma_2)^\dagger(\sigma_\mu^-(D_L)[\tfrac{1}{\sqrt{2}}q_1 g W^{+\mu}]\Gamma_2)cos(\beta)$$

$$-(U_L\Gamma_2)^\dagger(\sigma_\mu^-(S_L)[\tfrac{1}{\sqrt{2}}q_1 g W^{+\mu}]\Gamma_2)sin(\beta)$$

indicates a transition of D_L via W^+ to U_L, proportional to $cos(\beta)$, and a transition of S_L via W^+ to U_L, proportional to $sin(\beta)$, this latter transition intergenerational. The angle β is just the Cabbibo angle, and the mixing is just conventional weak mixing. Leptons do not similarly mix because the extra $SU(2)$ phase does not appear in those cases (see 4.56, where $\rho_+\Psi\Delta_0$ and $\rho_+\Psi\Delta_1$ are lepton terms). As to righthanded fields, because they are free of $SU(2)$ charged currents, they are free of mixing as well.

Certainly this is not a derivation of the Cabbibo angle. It nicely explains why leptons are immune to weak mixing, but it doesn't explain the exact $SU(2)$ phase needed for this purpose. Nor does it explain why there should be another family, nor force the inclusion of a third. When I first began work with the tensored division algebras I assumed that the occurance in **T** of a family plus antifamily was an indication that eventually there would be found to be an even number of families. Experimental evidence seems to indicate there are only three. For the time being, therefore, I consider the generation structure of this model an open question, with the ideas presented in this section an indication of how things might work out for the first two generations.

4.6. Gauging $SU(3)$.

A set of eight generators for $SU(3)$ are:

$$\begin{aligned}
\lambda_1 &= \tfrac{1}{2}(e_{L26} - e_{L34}), \\
\lambda_2 &= \tfrac{1}{2}(e_{L45} - e_{L61}), \\
\lambda_3 &= \tfrac{1}{2}(e_{L56} - e_{L41}), \\
\lambda_4 &= \tfrac{1}{2}(e_{L13} - e_{L52}), \\
\lambda_5 &= \tfrac{1}{2}(e_{L63} - e_{L24}), \\
\lambda_6 &= \tfrac{1}{2}(e_{L35} - e_{L12}), \\
\lambda_7 &= \tfrac{1}{2}(e_{L23} - e_{L15}), \\
\lambda_8 &= \tfrac{1}{3}(-e_{L23} - e_{L15} + 2e_{L46})
\end{aligned} \tag{4.61}$$

(we'll let the context distinguish the λ-projection operators from these generators, which will only be employed in this section; this notation for $SU(3)$ generators is conventional, and originated with Gell-Mann). The last two are diagonal. We see this by operating the Hermitian generators $\frac{i}{2}\lambda_7$ and $\frac{i}{2}\lambda_8$ on $\rho_+ e_r, r = 1, 2, 4$ (since

$$\rho_+\Psi = \rho_+(\Psi_+^0 + \Psi_+^1 e_1 + \Psi_+^2 e_2 + \Psi_+^4 e_4),$$

and we have identified the $\rho_+ e_r$, $r = 1, 2, 4$, directions as red-green-blue). For example,

$$\begin{aligned}
\tfrac{i}{2}\lambda_7[\rho_+ e_1] &= \tfrac{i}{4}\rho_+(e_2(e_6 e_1) - e_1(e_5 e_1)) \\
&= -\tfrac{i}{2}\rho_+ e_5 = \tfrac{i}{2}\rho_+ i e_1 \\
&= -\tfrac{1}{2}[\rho_+ e_1],
\end{aligned}$$

and

$$\begin{aligned}\tfrac{i}{2}\lambda_8[\rho_+e_1] &= \tfrac{i}{6}\rho_+(-e_2(e_6e_1) - e_1(e_5e_1) + 2e_4(e_6e_1)) \\ &= -\tfrac{i}{3}\rho_+e_5 = \tfrac{i}{3}\rho_+ie_1 \\ &= -\tfrac{1}{3}[\rho_+e_1].\end{aligned}$$

In like manner we determine a complete table of $SU(3)$ charges for $\rho_+e_r, r = 1,2,4$:

$$\begin{aligned}&\tfrac{i}{2}\lambda_7[\rho_+e_1] = -\tfrac{1}{2}[\rho_+e_1] \quad \text{and} \quad \tfrac{i}{2}\lambda_8[\rho_+e_1] = -\tfrac{1}{3}[\rho_+e_1], \\ &\tfrac{i}{2}\lambda_7[\rho_+e_2] = +\tfrac{1}{2}[\rho_+e_2] \quad \text{and} \quad \tfrac{i}{2}\lambda_8[\rho_+e_2] = -\tfrac{1}{3}[\rho_+e_2], \\ &\tfrac{i}{2}\lambda_7[\rho_+e_3] = 0[\rho_+e_3] \quad \text{and} \quad \tfrac{i}{2}\lambda_8[\rho_+e_3] = +\tfrac{2}{3}[\rho_+e_3].\end{aligned} \tag{4.62}$$

These quantum numbers justify the red-green-blue labels above, and the use of the Gell-Mann symbols $\lambda_a, a = 1, ..., 8$, to signify the generators. The antimatter charges of $\rho_-\Psi$ are opposite those above (because $ie_7\rho_\pm = \pm\rho_\pm$).

$SU(3)$ gauge fields arise from the (1,3)-matter Lagrangian 4.38 as follows:

$$\boxed{\mathcal{L}_{SU(3)} = constant < \rho_+\Psi, \gamma_\mu G^{\mu a}\lambda_a[\rho_+\Psi] >,} \tag{4.63}$$

where the $G^{\mu a}$ are the gluon fields ($\mu = 0,1,2,3,\ a = 1, ..., 8$). Once again we can obtain from those parts of $\mathcal{L}_{SU(3)}$ that can contribute a nonzero part to the result of the inner product 4.63 a list of possible transitions. For example, consider the simpler looking term

$$\begin{aligned}(\rho_+e_4)^\dagger(\lambda_5[\rho_+e_2]) &= -\frac{1}{2}(e_4\rho_+)(\rho_+(e_6(e_3e_2) - e_2(e_4e_2))) \\ &= (e_4\rho_+)(\rho_+e_4) \\ &= e_4\rho_+\rho_+e_4 \\ &= e_4\rho_+e_4 \\ &= \rho_-(e_4)^2 \\ &= -\rho_- \\ &= -\frac{1}{2} + \frac{1}{2}ie_7\end{aligned} \tag{4.64}$$

(the nonassociativity of **O** plays no role after the second line), which has a nonzero real part. The occurance of this expression in $\mathcal{L}_{SU(3)}$, together with its contributing a nonzero real value to the inner product, tells us that the green quark, $\rho_+ e_2$, can interact with $\gamma_\mu G^{\mu 5}\lambda_5$ to produce a blue quark, $\rho_+ e_4$.

In addition,

$$\begin{aligned}
(\rho_+ e_2)^\dagger(\lambda_5[\rho_+ e_4]) \quad &= -\frac{1}{2}(e_2\rho_+)(\rho_+(e_6(e_3e_4) - e_2(e_4e_4))) \\
&= -(e_2\rho_+)(\rho_+ e_2) \\
&= -e_2\rho_+\rho_+ e_2 \\
&= -e_2\rho_+ e_2 \\
&= -\rho_-(e_2)^2 \\
&= \rho_- \\
&= \frac{1}{2} - \frac{1}{2} i e_7,
\end{aligned}$$

which again has a nonzero real part, so with the help of $\gamma_\mu G^{\mu 5}\lambda_5$, a blue quark, $\rho_+ e_4$, can transform to a green quark, $\rho_+ e_2$.

However,

$$\begin{aligned}
(\rho_+ e_1)^\dagger(\lambda_5[\rho_+ e_2]) \quad &= (e_1\rho_+)(\rho_+ e_4) \\
&= ((e_1\rho_+)\rho_-)e_4 \\
&= (e_1(\rho_+\rho_-))e_4 \\
&= 0
\end{aligned}$$

(nonassociativity is responsible for $\rho_+ \to \rho_-$ in the second line), so the green quark, $\rho_+ e_2$, can not interact with $\gamma_\mu G^{\mu 5}\lambda_5$ to produce a red quark, $\rho_+ e_1$.

On the other hand, although

$$\begin{aligned}
(\rho_- e_1)^\dagger(\lambda_5[\rho_+ e_2]) \quad &= (e_1\rho_-)(\rho_+ e_4) \\
&= ((e_1\rho_-)\rho_-)e_4 \\
&= (e_1\rho_-)e_4 \\
&= (\rho_+ e_1)e_4 \\
&= \rho_-(e_1e_4) \\
&= -\rho_- e_3 \\
&\neq 0,
\end{aligned}$$

it has no real part, hence drops out of the inner product, so the green quark, $\rho_+ e_2$, can not interact with $\gamma_\mu G^{\mu 5}\lambda_5$ to produce an antired antiquark, $\rho_- e_1$, either.

Note that the space index μ in 4.63 appears in γ_μ, not just σ_μ, and that Ψ is the full left-right spinor hyperfield. That is, parity is conserved in this case, which is a direct consequence of the $SU(3)$ invariance of the Γ_m, in contrast to $SU(2)$. In the next chapter we shall see that $SU(3)$ is also necessarily exact.

Also note that index doubling takes $e_1 \to e_2 \to e_4 \to e_1$, and makes corresponding changes in the λ_a. Index quadrupalling takes $e_1 \to e_4 \to e_2 \to e_1$, with yet another set of λ_a. Doubling again gets us back where we started. One might conjecture that these three representations are somehow linked to there being three generations, but this is admittedly numerological groping, and it will not be pursued here. In fact, please forget I even mentioned it. There are more likely candidate explanations for family replication.

5. Spontaneous Symmetry Breaking.

5.1. Scalar Fields.

Of the positive attributes of model building with $\mathbf{T}$, rigidity ranks high on the list. The ability to mirror the structural features of our physical reality with flexible mathematics bespeaks more the talents of the model builder than any fundamental connection of the mathematics to reality. To lay claim to such a connection a mathematical idea must, in my opinion, satisfy at least the three qualities that so distinguish $\mathbf{T}$: selectness; inflexibility; and consistency with those aspects of physical reality it is intended to explicate. This chapter should highlight these qualities.

Two kinds of fields have already been defined on this developing context: the fermion hyperfield, Ψ; and the gauge fields B^{μ}, $A^{\mu j}$, and $G^{\mu a}$. The former resides in $\mathbf{T}^2$, the spinor/object space for the external, or geometric algebra $\mathbf{T}_L(2)$, and also for the internal algebra $\mathbf{P}_R$. The latter fields reside in, or are associated with, these adjoint algebras. The former are in some sense primary, the latter derived or secondary. The inner product, $< ..., ... >$, is defined for primary fields, elements of $\mathbf{T}$ or $\mathbf{T}^2$, and in what follows we shall be considering scalar Lagrangians of the form

$$\mathcal{L}_\phi = < \partial_\mu \phi, \partial^\mu \phi > - \mu^2 < \phi, \phi > - \lambda < \phi, \phi >^2, \tag{5.1}$$

where ϕ is a scalar field. Because our ϕ is not a gauge field, and because we shall be required to take its inner product, it is defined here to be a primary field. This, as it turns out, is a nontrivial constraint.

We may define a field to be scalar with respect to a given geometry if its algebraic action on a spinor field commutes with the action of the geometric (Clifford) algebra of the geometry on the same spinor. For example, let ϕ be a scalar with respect to $\mathbf{R}_{1,9} \subseteq \mathbf{T}_L(2)$. In this case we may assume $\phi \in \mathbf{P}$, and that its algebraic connection to Ψ takes the form,

$$\Psi\phi. \tag{5.2}$$

Note that as

$$\mathbf{O}_L = \mathbf{O}_R \subseteq \mathbf{R}_{1,9},$$

ϕ must be free of hypercomplex octonions, and if dependent on hypercomplex quaternions it must connect to Ψ from the right to commute with $\mathbf{Q}_L \subseteq \mathbf{R}_{1,9}$.

A second possibility is to consider scalars with respect to the subalgebras $\rho_{L\pm}\mathbf{T}_L(2)\rho_{L\pm}$ (see 4.7), the matter and antimatter projections from $\mathbf{T}_L(2)$. Multiplication from the left by e_7 on Ψ commutes with these algebras, so the dimension of the scalar space is doubled, and we may define a new action on Ψ by,

$$\rho_+\Psi\phi_+ + \rho_-\Psi\phi_-, \tag{5.3}$$

where $\phi_\pm \in \mathbf{P}$.

[Note that although $\mathbf{O}_L = \mathbf{O}_R$, $e_{L7} \neq e_{R7}$, and in fact e_{R7} does not commute with $\rho_{L\pm}\mathbf{T}_L(2)\rho_{L\pm}$, even though e_{L7} does. For example,

$$\rho_{L+}e_{L12}\epsilon \in \rho_{L+}\mathbf{T}_L(2)\rho_{L+},$$

but

$$\begin{aligned}
\rho_{L+}e_{L12}e_{R7}[\Psi] &= \tfrac{1}{2}\rho_{L+}e_{L12}(-e_{L7} + e_{L15} + e_{L23} + e_{L46})[\Psi] \\
&= \tfrac{1}{2}(-e_{L7} - e_{L15} - e_{L23} + e_{L46})\rho_{L+}e_{L12}[\Psi] \\
&= -e_{R46}\rho_{L+}e_{L12}[\Psi] \\
&\neq e_{R7}\rho_{L+}e_{L12}[\Psi].
\end{aligned}$$

Hence the commutator

$$[\rho_{L+}e_{L12}, e_{R7}] \neq 0,$$

so e_{R7} does not commute with $\rho_{L+}\mathbf{T}_L(2)\rho_{L+}$.]

This scalar has the advantage that it distinguishes matter from antimatter, but it has the problem that in giving masses to leptons (via spontaneous symmetry breaking), it is unable to do the same for the quarks, which are not diagonal with respect to the resulting mass operator.

Consequently we shall loosen the definition one final step (the last possible) and take our scalars with respect to the matter and antimatter Dirac algebras, $\mathbf{R}_{1,3}\rho_{L\pm}$ ($\mathbf{R}_{1,3}\rho_{L+}$ is the boxed subalgebra in 4.7). These algebras commute with e_{R7} as well as e_{L7}, which again doubles the dimension of our algebraic scalar space. In this case we define the connection to Ψ by

$$\boxed{\rho_+\Psi\rho_+\phi_0 + \rho_+\Psi\rho_-\phi_1 + \rho_-\Psi\rho_+\phi_2 + \rho_-\Psi\rho_-\phi_3,} \tag{5.4}$$

where $\phi_m \in \mathbf{P}$, m=0,1,2,3.

Recall that we have made the following identifications:

$$\rho_+\Psi\rho_+ = \text{ lepton doublet;}$$

$$\rho_+\Psi\rho_- = \text{ quark doublet, color triplet;}$$

$\rho_-\Psi\rho_-$ and $\rho_-\Psi\rho_+$ respective antimatter hyperfields. Each of the ϕ_m is an element of $\mathbf{P}$, hence $SU(3)$ singlets, so they mix the doublet parts of these hyperfields, but not the $SU(3)$ triplet parts. *Consequently spontaneous symmetry breaking can occur only to the $SU(2)$ level. $SU(3)$ is necessarily exact.* As was the case with parity nonconservation (see section 4.3), the exactness of $SU(3)$, and the breaking of $SU(2)$, follow from the mathematics.

5.2. Scalar Lagrangians.

Let $\phi \in \mathbf{P}$ be one of the scalar fields defined in 5.4. Define

$$\boxed{\mathcal{L}_\phi = \sum_{\nu=0}^{3} < \partial_\nu\phi, \partial^\nu\phi > -\mu^2 < \phi, \phi > -\lambda < \phi, \phi >^2 .} \tag{5.5}$$

This is very nearly as conventional a scalar Lagrangian as it looks, even though ϕ in its present form is not an $SU(2)$ doublet, or even a pair of doublets. If

$$e^{\tilde{\alpha}/2} \in SU(2)$$

(see 3.55 and 4.13), it is natural to define the $SU(2)$ transformation of ϕ by:

$$\boxed{\phi \to e^{\tilde{\alpha}/2}\phi e^{-\tilde{\alpha}/2}} \tag{5.6}$$

(so ϕ would be $U(1)$ invariant)). ϕ therefore transforms like the direct sum of an $SU(2)$ singlet and triplet.

However,

$$\boxed{<\partial_\mu\phi,\partial^\mu\phi>=\tfrac{1}{8}\{[\textstyle\sum_{mn}M_m((\partial_\mu\phi\Gamma_n)^\dagger(\partial^\mu\phi\Gamma_n))]+[same]^\dagger\},} \quad (5.7)$$

and under $SU(2)$,

$$\boxed{\phi\Gamma_n \rightarrow e^{\tilde{\alpha}/2}\phi e^{-\tilde{\alpha}/2}e^{\tilde{\alpha}/2}\Gamma_n = e^{\tilde{\alpha}/2}\phi\Gamma_n.} \quad (5.8)$$

That is, $\phi\Gamma_n$ is an $SU(2)$ doublet, and $SU(2)$ gauge fields arise in 5.7 consistent with 5.8, not 5.6, because the derivatives act on $\phi\Gamma_n$. (Confession: were the directions from which the partials act in 5.5 reversed, gauge fields would arise consistent with 5.6 instead of 5.8, and the result would be less conventional. I'm pretending that this is for some reason precluded.)

Gauging $U(1)$ is slightly more complicated. Because the ϕ are $U(1)$ invariant ($e^{\theta e_7}\phi e^{-\theta e_7} = e^{i\theta}\phi e^{-i\theta} = \phi$), the $U(1)$ charge of $\phi\Gamma_n$ derives entirely from Γ_n (see 3.62). Consequently there is no single covariant derivative that can replace ∂^μ in 5.7. Instead, $U(2)$ gauge fields arise as follows:

$$\begin{aligned}
\partial^\mu\phi\Gamma_n &\rightarrow (\partial^\mu - \tfrac{1}{2}g'iB^\mu - \tfrac{1}{2}gA^{\mu j}\tau_j)\phi\Gamma_n, \quad n=0,2,\\
\partial^\mu\phi\Gamma_n &\rightarrow (\partial^\mu + \tfrac{1}{2}g'iB^\mu - \tfrac{1}{2}gA^{\mu j}\tau_j)\phi\Gamma_n, \quad n=1,3
\end{aligned} \quad (5.9)$$

(the τ_j are defined in 4.40). To simplify what follows, note that

$$\begin{aligned}
[-\tfrac{1}{2}g'iB^\mu - \tfrac{1}{2}gA^{\mu j}\tau_j] &= (\lambda_+ + \lambda_-)[...](\lambda_+ + \lambda_-)\\
&= \lambda_+[...]\lambda_+ + \lambda_+[...]\lambda_- + \lambda_-[...]\lambda_+ + \lambda_-[...]\lambda_-\\
&= (-\tfrac{1}{2}g'iB^\mu + \tfrac{1}{2}giA^{\mu 3})\lambda_+ - (\tfrac{1}{2}g(A^{\mu 1} + iA^{\mu 2}))q_1\lambda_-\\
&\quad -(\tfrac{1}{2}g(A^{\mu 1} - iA^{\mu 2}))q_1\lambda_+ + (-\tfrac{1}{2}g'iB^\mu - \tfrac{1}{2}giA^{\mu 3})\lambda_-\\
&= -[ig_{z1}Z^\mu]\lambda_+ - [\tfrac{g}{2^{1/2}}W^{\mu -}]q_1\lambda_-\\
&\quad -[\tfrac{g}{2^{1/2}}W^{\mu +}]q_1\lambda_+ - [i(eA^\mu + g_{z2}Z^\mu)]\lambda_-,
\end{aligned} \quad (5.10)$$

where

$$g_{z1} = (g^2 + g'^2)^{1/2}/2^{1/2},$$

and

$$g_{z2} = (g'^2 - g^2)/(2(g^2 + g'^2)^{1/2}).$$

Likewise,

$$\begin{aligned}[+\tfrac{1}{2}g'iB^\mu - \tfrac{1}{2}gA^{\mu j}\tau_j] &= [i(eA^\mu + g_{z2}Z^\mu)]\lambda_+ - [\tfrac{g}{\sqrt{2}}W^{\mu -}]q_1\lambda_- \\ &\quad -[\tfrac{g}{\sqrt{2}}W^{\mu +}]q_1\lambda_+ + [ig_{z1}Z^\mu]\lambda_-.\end{aligned} \tag{5.11}$$

Recall that prior to $U(2)$ rotations, $\Gamma_0 = \lambda_+\rho_+$, $\Gamma_1 = \lambda_-\rho_+$, $\Gamma_2 = \lambda_+\rho_-$, $\Gamma_3 = \lambda_-\rho_-$ (unrotated, $\Gamma_m = \Delta_m$), and that $\lambda_+\lambda_- = \lambda_-\lambda_+ = 0$, $\lambda_\pm^2 = \lambda_\pm$. Also, $q_1\lambda_\pm = \lambda_\mp q_1$. Decompose

$$\begin{aligned}\phi &= \lambda_+\phi\lambda_+ + \lambda_+\phi\lambda_- + \lambda_-\phi\lambda_+ + \lambda_-\phi\lambda_- \\ &= \phi_{++}\lambda_+ + \phi_{+-}q_1\lambda_- + \phi_{-+}q_1\lambda_+ + \phi_{--}\lambda_- \\ &= \lambda_+\phi_{++} + \lambda_+q_1\phi_{+-} + \lambda_-q_1\phi_{-+} + \lambda_-\phi_{--},\end{aligned} \tag{5.12}$$

where $\phi_{++}, \phi_{+-}, \phi_{-+}, \phi_{--} \in \mathbf{C}$. Therefore, for n=0,2,

$$\begin{aligned}[-\tfrac{1}{2}g'iB^\mu - \tfrac{1}{2}gA^{\mu j}\tau_j]\phi\Gamma_n &= [...](\phi_{++}\lambda_+ + \phi_{-+}q_1\lambda_+)\Gamma_n \\ &= [...](\lambda_+\phi_{++} + \lambda_-q_1\phi_{-+})\Gamma_n \\ &= \{-ig_{z1}Z^\mu - \tfrac{g}{\sqrt{2}}W^{\mu +}q_1\}\phi_{++}\Gamma_n \\ &\quad +\{\tfrac{g}{\sqrt{2}}W^{\mu -} - i(eA^\mu + g_{z2}Z^\mu)q_1\}\phi_{-+}\Gamma_n,\end{aligned} \tag{5.13}$$

where the $\lambda_\pm$ have been absorbed into Γ_n in the last step. Likewise, for n=1,3,

$$\begin{aligned}[+\tfrac{1}{2}g'iB^\mu - \tfrac{1}{2}gA^{\mu j}\tau_j]\phi\Gamma_n &= [...](\lambda_+q_1\phi_{+-} + \lambda_-\phi_{--})\Gamma_n \\ &= \{i(eA^\mu + g_{z2}Z^\mu)q_1 + \tfrac{g}{\sqrt{2}}W^{\mu +}\}\phi_{+-}\Gamma_n \\ &\quad +\{-\tfrac{g}{\sqrt{2}}W^{\mu -}q_1 - ig_{z1}Z^\mu\}\phi_{--}\Gamma_n.\end{aligned} \tag{5.14}$$

As usual, when μ^2 in 5.5 is taken negative, ϕ takes on a nonzero vacuum expectation value (vev) **[12]**, in this case somewhere on the 7-sphere

$$< \phi, \phi >= \text{constant} \tag{5.15}$$

(which is the equation of a 7-sphere). Because A^μ is identified with the massless photon field, the vacuum expectation values of ϕ_{-+}, associated with A^μ in 5.13, and ϕ_{+-}, associated with A^μ in 5.14, must be zero. ϕ_{-+} and ϕ_{+-} are the charged parts of ϕ.

Let ν_+ and ν_- be the vev's of ϕ_{++} and ϕ_{--} (the neutral bits of ϕ). Including only these chargeless parts of ϕ, 5.13 and 5.14 reduce to (n=0,2)

$$[-\tfrac{1}{2}g'iB^\mu - \tfrac{1}{2}gA^{\mu j}\tau_j]\phi\Gamma_n \quad\rightarrow\quad \{-ig_{z1}Z^\mu - \tfrac{g}{\sqrt{2}}W^{\mu+}\}\nu_+\Gamma_n, \tag{5.16}$$

and (n=1,3)

$$[+\tfrac{1}{2}g'iB^\mu - \tfrac{1}{2}gA^{\mu j}\tau_j]\phi\Gamma_n \quad\rightarrow\quad \{-ig_{z1}Z^\mu - \tfrac{g}{\sqrt{2}}W^{\mu-}\}\nu_-\Gamma_n. \tag{5.17}$$

The resulting mass terms in 5.5 are:

$$\tfrac{1}{8}\{[\textstyle\sum_m M_m(((-ig_{z1}Z_\mu - \tfrac{g}{\sqrt{2}}W_\mu^+ q_1)\nu_{++}(\Gamma_0+\Gamma_2))^\dagger(\text{same}, \mu\text{ up}))$$

$$+\textstyle\sum_m M_m(((-ig_{z1}Z_\mu - \tfrac{g}{\sqrt{2}}W_\mu^- q_1)\nu_{--}(\Gamma_1+\Gamma_3))^\dagger(\text{same}, \mu\text{ up}))] + [\text{same}]^\dagger\}$$

(assuming Γ_m are rotated to their canonical values Δ_m, so that $\Gamma_0+\Gamma_2=\lambda_+, \Gamma_1+\Gamma_3=\lambda_-$)

$$= \tfrac{1}{8}\{[\textstyle\sum_m M_m(\lambda_+(ig_{z1}Z_\mu^\dagger + \tfrac{g}{\sqrt{2}}(W_\mu^+)^\dagger q_1)((-ig_{z1}Z_\mu - \tfrac{g}{\sqrt{2}}(W_\mu^+)q_1)\lambda_+\nu_{++}^2)$$

$$+\textstyle\sum_m M_m(\lambda_-(ig_{z1}Z_\mu^\dagger + \tfrac{g}{\sqrt{2}}(W_\mu^-)^\dagger q_1)((-ig_{z1}Z_\mu - \tfrac{g}{\sqrt{2}}(W_\mu^-)q_1)\lambda_-\nu_{--}^2)] + [\text{sar}$$

$$= \tfrac{1}{8}\{[\textstyle\sum_m M_m(\lambda_+(g_{z1}^2 Z_\mu Z^\mu + \tfrac{g^2}{2}(W_\mu^+)^\dagger W^{\mu+})\nu_{++}^2$$

$$+\textstyle\sum_m M_m(\lambda_-(g_{z1}^2 Z_\mu Z^\mu + \tfrac{g^2}{2}(W_\mu^-)^\dagger W^{\mu-})\nu_{--}^2] + [\text{same}]^\dagger\}$$

$$= \tfrac{1}{8}\{[2((g_{z1}^2 Z_\mu Z^\mu + \tfrac{g^2}{2}(W_\mu^+)^\dagger W^{\mu+})\nu_{++}^2$$

$$+2((g_{z1}^2 Z_\mu Z^\mu + \tfrac{g^2}{2}(W_\mu^-)^\dagger W^{\mu-})\nu_{--}^2] + [\text{same}]^\dagger\}$$

$$= \tfrac{\nu^2}{8}((g^2+g'^2)Z_\mu Z^\mu + g^2(W_\mu^+)^\dagger W^{\mu+} + g^2 W_\mu^-)^\dagger W^{\mu-}), \tag{5.18}$$

where $\nu^2 = \nu_{++}^2 + \nu_{--}^2$. Therefore, as expected (θ_w = Weinberg angle),

$$\frac{M_W}{M_Z} = \frac{g}{(g^2+g'^2)^{1/2}} = cos\theta_w. \tag{5.19}$$

5.3. Fermions and Scalars.

In 5.4 four scalar fields were defined, ϕ_m, m=0,1,2,3, along with their links to the fermion hyperfield Ψ. Including only the nonzero vev's of each, these connections reduce to:

$$\begin{aligned} &\rho_+\Psi\rho_+(\nu_{0+}\lambda_+ + \nu_{0-}\lambda_-) + \rho_+\Psi\rho_-(\nu_{1+}\lambda_+ + \nu_{1-}\lambda_-) + ... \\ &= \rho_+\Psi\Delta_0\nu_{0+} + \rho_+\Psi\Delta_1\nu_{0-} + \rho_+\Psi\Delta_2\nu_{1+} + \rho_+\Psi\Delta_3\nu_{1-} + ... \end{aligned} \quad (5.20)$$

Consider the inner product of this with Ψ, and in particular the term

$$\begin{aligned} (\Psi\Gamma_m)\hat{}((\rho_+\Psi\Delta_0\nu_{0+})\Gamma_m) &= \nu_{0+}[(\Psi\Gamma_m)\hat{}(\rho_+\Psi\Delta_0\Gamma_m)] \\ &\rightarrow \nu_{0+}[(\Psi\Gamma_0)\hat{}(\rho_+\Psi\Gamma_0)] \\ &\rightarrow \nu_{0+}[(\rho_+\Psi\Gamma_0)\hat{}(\rho_+\Psi\Gamma_0)] \\ &= \nu_{0+}N\hat{}N, \end{aligned} \quad (5.21)$$

where in line 2 it is assumed that Γ_0 is unrotated, ie., $\Gamma_0 = \Delta_0$, and in line 3 just the part of line 2 that survives the inner product is indicated, which in the last step is seen to imply a mass for the neutrino proportional to ν_{0+}. Following these same steps for the other terms yields:

$$\begin{array}{rcl} M_{neutrino} & \sim & \nu_{0+} \\ M_{lepton(-1)} & \sim & \nu_{0-} \\ M_{quark(2/3)} & \sim & \nu_{1+} \\ M_{quark(-1/3)} & \sim & \nu_{1-} \\ M_{antiquark(1/3)} & \sim & \nu_{2+} \\ M_{antiquark(-2/3)} & \sim & \nu_{2-} \\ M_{antilepton(+1)} & \sim & \nu_{3+} \\ M_{antineutrino} & \sim & \nu_{3-} \end{array} \quad (5.22)$$

(see 4.17 and 4.19; $\sim$ means "proportional to" here).

The mass proportions indicated in 5.22 are for Dirac masses, but no exact values are determined. It is now only worth pointing out that neutrinos have Dirac masses, and there is sufficient freedom to distinguish the family and antifamily at the mass level.

6. 10 Dimensions.

6.1. Fermion Lagrangian.

There is every reason to believe that the (1,3)-Minkowski geometry is secondary. Our model building is not based upon the Pauli algebra, $\mathbf{P}_L$, from which is made the Dirac algebra for $\mathbf{R}^{1,3}$, but upon $\mathbf{T}_L$, which bears a similar relationship to $\mathbf{R}^{1,9}$ (see 4.2 and 4.4). Define

$$\tilde{\sigma}_0 = 1, \quad \tilde{\sigma}_j = -e_{L7}q_j, j = 1,2,3, \quad \tilde{\sigma}_{p+3} = -ie_{Lp}, p = 1,...,6. \tag{6.1}$$

These are related to the Clifford algebra $\mathbf{R}_{1,9}$ given in 4.4 as the elements

$$\sigma_0 = 1, \; \sigma_j = iq_j, j = 1,2,3,$$

are related to the Clifford algebra $\mathbf{R}_{1,3}$ given in 4.2. We should require, therefore, that a field theory founded upon $\mathbf{T}_L$ and $\mathbf{T}$ should involve functions of $\mathbf{R}^{1,9}$, not just $\mathbf{R}^{1,3}$.

We shall consequently need to replace the (1,3)-Dirac operator, $\partial\!\!\!/_+$, employed in 4.38 in the definition of $\mathcal{L}_0$, with a (1,9)-Dirac operator. To ensure that it violate parity as does $\partial\!\!\!/_+$, we define

$$\boxed{\partial\!\!\!/_{1,9} = \begin{bmatrix} 0 & \sum_0^9 \tilde{\sigma}_a \partial^a \\ \sum_0^9 \tilde{\sigma}_a^- \overleftarrow{\partial^a} & 0 \end{bmatrix}}, \tag{6.2}$$

where

$$\tilde{\sigma}_0^- = \tilde{\sigma}_0, \quad \tilde{\sigma}_a^- = -\tilde{\sigma}_a, a = 1,...,9. \tag{6.3}$$

Also define

$$\boxed{\partial\!\!\!/_{1,3} = \begin{bmatrix} 0 & \sum_0^3 \tilde{\sigma}_\mu \partial^\mu \\ \sum_0^3 \tilde{\sigma}_\mu^- \overleftarrow{\partial^\mu} & 0 \end{bmatrix}}, \tag{6.4}$$

and

$$\boxed{\partial\!\!\!/_{0,6} = \partial\!\!\!/_{1,9} - \partial\!\!\!/_{1,3}.} \tag{6.5}$$

Note that

$$\begin{aligned} \rho_{L+}\partial\!\!\!/_{1,9}\rho_{L+} &= \rho_{L+}\partial\!\!\!/_{1,3}\rho_{L+} + \rho_{L+}\partial\!\!\!/_{0,6}\rho_{L+} \\ &= \rho_{L+}\partial\!\!\!/_{1,3}\rho_{L+} \\ &= \partial\!\!\!/_{+}. \end{aligned}$$

The second term drops out because

$$\rho_{L+}e_{Lp}\rho_{L+} = \rho_{L+}\rho_{L-}e_{Lp} = 0,\ p = 1, ..., 6;$$

so $\rho_{L+}\partial\!\!\!/_{0,6}\rho_{L+} = 0$. Also, because $\rho_{L+}e_{L7}\rho_{L+} = -i\rho_{L+}$,

$$\rho_{L+}\tilde{\sigma}_\mu\rho_{L+} = \sigma_\mu\rho_{L+},\ \mu = 0, 1, 2, 3,$$

so

$$\rho_{L+}\partial\!\!\!/_{1,3}\rho_{L+} = \partial\!\!\!/_{1,3}\rho_{L+} = \partial\!\!\!/_{+}.$$

(In replacing ρ_{L+} by ρ_{L-}, space components are reflected.)

Assume now that the fermion hyperfield Ψ is a function of $\mathbf{R}^{1,9}$ coordinates, and define a new fermion Lagrangian:

$$\boxed{\mathcal{L}_{1,9} =< \Psi, \partial\!\!\!/_{1,9}\Psi > .} \tag{6.6}$$

Since $\partial\!\!\!/_{1,9} = \partial\!\!\!/_{1,3} + \partial\!\!\!/_{0,6}$, and $\rho_{L+} + \rho_{L-} = 1$,

$$\begin{aligned} \partial\!\!\!/_{1,9} &= \partial\!\!\!/_{1,3}\rho_{L+} + \partial\!\!\!/_{1,3}\rho_{L-} + \partial\!\!\!/_{0,6}\rho_{L+} + \partial\!\!\!/_{0,6}\rho_{L-} \\ &= \partial\!\!\!/_{+}\rho_{L+} + \partial\!\!\!/_{-}\rho_{L-} + \partial\!\!\!/_{0,6}\rho_{L+} + \partial\!\!\!/_{0,6}\rho_{L-}, \end{aligned} \tag{6.7}$$

where $\partial\!\!\!/_{-}$ is the antimatter version of $\partial\!\!\!/_{+}$. Note that $\partial\!\!\!/_{\pm}\rho_{L\pm} = \partial\!\!\!/_{\pm}$; the projection operators $\rho_{L\pm}$ can be absorbed into $\partial\!\!\!/_{\pm}$, but they are left in place in 6.7 to clarify what follows.

First note that if $X, Y \in \mathbf{T}$, then

$$< X, \rho_{+}Y >=< \rho_{+}X, \rho_{+}Y > . \tag{6.8}$$

That is,

$$<\rho_-X,\rho_+Y>=0 \tag{6.9}$$

in general. Even more generally,

$$<X,Y>=<\rho_+X,\rho_+Y>+<\rho_-X,\rho_-Y>.$$

To prove this we need only prove that for $a,b\in\{0,...,7\}$, the expression $(e_a\rho_-)(\rho_+e_b)$ can have no nonzero real part. It is not hard to see that if e_a, e_b, and e_7 do not associate (that is, $(e_ae_b)e_7=-e_a(e_be_7)$), then

$$\begin{aligned}(e_a\rho_-)(\rho_+e_b) &= e_a(\rho_+(\rho_+e_b))\\ &= e_a(\rho_+e_b)\\ &= e_a(e_b\rho_-)\\ &= (e_ae_b)\rho_+.\end{aligned}$$

Since e_a, e_b, and e_7 do not associate, $e_ae_b=\pm e_p$ for some $p=1,...,6$, hence $e_ae_b)\rho_+$ has no real part. If, on the other hand, $(e_ae_b)e_7=e_a(e_be_7)$, then

$$(e_a\rho_-)(\rho_+e_b)=e_a\rho_-\rho_+e_b=0,$$

since $\rho_-\rho_+=0$. This proves 6.8 and 6.9.

Therefore, noting that

$$\begin{aligned}\not{\partial}_{0,6}\rho_{L\pm} &= \rho_{L\mp}\not{\partial}_{0,6}\rho_{L\pm},\\ \not{\partial}_{\pm}\rho_{L\pm} &= \rho_{L\pm}\not{\partial}_{\pm}\rho_{L\pm},\end{aligned}$$

we have

$$\begin{aligned}\boxed{\mathcal{L}_{1,9}=} \quad & <\Psi,\not{\partial}_+\rho_{L+}\Psi>+<\Psi,\not{\partial}_-\rho_{L-}\Psi>+<\Psi,\not{\partial}_{0,6}\rho_{L+}\Psi>\\ & +<\Psi,\not{\partial}_{0,6}\rho_{L-}\Psi>\\ = \quad & <\rho_+\Psi,\rho_+(\not{\partial}_+(\rho_+\Psi))>+<\rho_-\Psi,\rho_-(\not{\partial}_-(\rho_-\Psi))>\\ & +<\rho_-\Psi,\rho_-(\not{\partial}_{0,6}(\rho_+\Psi))>+<\rho_+\Psi,\rho_+(\not{\partial}_{0,6}(\rho_-\Psi))>\\ = \quad & <(\rho_+\Psi),\not{\partial}_+(\rho_+\Psi)>+<(\rho_-\Psi),\not{\partial}_-(\rho_-\Psi)>\\ & +<(\rho_-\Psi),\not{\partial}_{0,6}(\rho_+\Psi)>+<(\rho_+\Psi),\not{\partial}_{0,6}(\rho_-\Psi)>.\end{aligned} \tag{6.10}$$

The first equality follows from 6.8 and 6.9. The first term after the last equality is just

$$< (\rho_+\Psi), \not{\partial}_+(\rho_{L+}\Psi) >=< \Psi, \not{\partial}_+\Psi >= \mathcal{L}_0, \tag{6.11}$$

the fermion Lagrangian studied in chapter 4. The second term is the antimatter version of $\mathcal{L}_0$. Both, as we have seen, give rise to reasonably conventional leptoquark gauge models, matching the Standard Model in most of their salient features. The former takes matter $(\rho_+\Psi)$ to matter $(\rho_+\Psi)$; the latter, antimatter $(\rho_-\Psi)$ to antimatter $(\rho_-\Psi)$.

MATTER/ANTIMATTER MIXING

The last two terms of 6.10 are new, and they are a problem. The first takes matter $(\rho_+\Psi)$ to antimatter $(\rho_-\Psi)$, the second the reverse. Such transitions are not observed, but they disappear from $\mathcal{L}_{1,9}$ if

$$\boxed{\not{\partial}_{0,6}(\rho_\pm\Psi) = 0} \tag{6.12}$$

identically. The conditions for this are developed below.

First we'll look at the term $\mathcal{L}_{-+} =< (\rho_-\Psi), \not{\partial}_{0,6}(\rho_+\Psi) >$ in more detail. Recall that we may make the decomposition,

$$\rho_\pm\Psi = \rho_\pm(\Psi^0_\pm + \Psi^1_\pm e_1 + \Psi^2_\pm e_2 + \Psi^4_\pm e_4), \tag{6.13}$$

where $\Psi^m_\mp$, m=0,1,2,4 are in $\mathbf{P}^2$. In particular, we designated the matter fields Ψ^m_+, m=0,1,2,4, lepton, red-, green-, and blue-quark. Consider the expression

$$\not{\partial}_{0,6}(\rho_+\Psi^1_+e_1)$$

which occurs in $\mathcal{L}_{-+}$. Because of nonassociativity,

$$e_p(\rho_+e_1) = (e_p\rho_+)e_1,\ p = 0, 1, 5, 7,$$

$$e_p(\rho_+e_1) = (e_p\rho_-)e_1,\ p = 2, 3, 4, 6.$$

Also recall,

$$\rho_\pm e_3 = \rho_\pm(e_7e_2) = \mp\rho_\pm ie_2,$$

$$\rho_\pm e_6 = \rho_\pm(e_7e_4) = \mp\rho_\pm ie_4,$$

$$\rho_\pm e_5 = \rho_\pm(e_7e_1) = \mp\rho_\pm ie_1,$$

Therefore,

$$\begin{aligned}
\partial\!\!\!/_{0,6}(\rho_+\Psi^1_+e_1) &= i\omega\textstyle\sum_1^6 e_{Lp}\partial^{p+3}[\rho_+\Psi^1_+e_1] \\
&= i\omega(\rho_-e_1\partial^4+\rho_+e_2\partial^5+\rho_+e_3\partial^6+\rho_+e_4\partial^7+\rho_-e_5\partial^8+\rho_+e_6\partial^9)[\Psi^1_+e_1] \\
&= i\omega(\rho_-e_1(\partial^4+i\partial^8)+\rho_+e_2(\partial^5-i\partial^6)+\rho_+e_4(\partial^7-i\partial^9))[\Psi^1_+e_1] \\
&= i\omega\rho_-[(e_1(\partial^4+i\partial^8)+e_2(\partial^5-i\partial^6)+e_4(\partial^7-i\partial^9))\Psi^1_+e_1].
\end{aligned}$$

Continuing along this line we can show

$$\boxed{\partial\!\!\!/_{0,6}(\rho_+\Psi_+) =}$$

$$\begin{aligned}
& i\omega(\rho_-e_1(\partial^4+i\partial^8)+\rho_-e_2(\partial^5+i\partial^6)+\rho_-e_4(\partial^7+i\partial^9))[\Psi^0_+] \\
& +i\omega(\rho_-e_1(\partial^4+i\partial^8)+\rho_+e_2(\partial^5-i\partial^6)+\rho_+e_4(\partial^7-i\partial^9))[\Psi^1_+e_1] \\
& +i\omega(\rho_+e_1(\partial^4-i\partial^8)+\rho_-e_2(\partial^5+i\partial^6)+\rho_+e_4(\partial^7-i\partial^9))[\Psi^2_+e_2] \\
& +i\omega(\rho_+e_1(\partial^4-i\partial^8)+\rho_+e_2(\partial^5-i\partial^6)+\rho_-e_4(\partial^7+i\partial^9))[\Psi^4_+e_4] \\
= {} & \rho_-i\omega[(\partial^4+i\partial^8)+e_2(\partial^5+i\partial^6)+e_4(\partial^7+i\partial^9))\Psi^0_+] \\
& +\rho_-i\omega[(e_1(\partial^4+i\partial^8)+e_2(\partial^5-i\partial^6)+e_4(\partial^7-i\partial^9))\Psi^1_+e_1] \\
& +\rho_-i\omega[(e_1(\partial^4-i\partial^8)+e_2(\partial^5+i\partial^6)+e_4(\partial^7-i\partial^9))\Psi^2_+e_2] \\
& +\rho_-i\omega[(e_1(\partial^4-i\partial^8)+e_2(\partial^5-i\partial^6)+e_4(\partial^7+i\partial^9))\Psi^4_+e_4] \\
= {} & \rho_-[\partial\!\!\!/_{0,6+++}\Psi^0_+ + \partial\!\!\!/_{0,6+--}\Psi^1_+e_1 + \partial\!\!\!/_{0,6-+-}\Psi^2_+e_2 + \partial\!\!\!/_{0,6--+}\Psi^4_+e_4].
\end{aligned} \tag{6.14}$$

Therefore $\mathcal{L}_{-+}$ may be written

$$\begin{aligned}
& <\rho_-\Psi, \partial\!\!\!/_{0,6}(\rho_+\Psi)> \\
& = <\rho_-(\Psi^0_- + \Psi^1_-e_1 + \Psi^2_-e_2 + \Psi^4_-e_4),\ \rho_-(i\omega(e_1(\partial^4+i\partial^8)\Psi^0_+ + ...)> .
\end{aligned} \tag{6.15}$$

Since, for example, $<\rho_-e_1, \rho_-e_1> \neq 0$, the last line above indicates a transition

$$\Psi^0_+ = \text{lepton} \rightarrow \text{antired-quark} = \Psi^1_-. \tag{6.16}$$

In fact, the last line in 6.15 gives rise to all possible transitions from fields with $SU(3)$ color white, red, green, and blue, to antifields of all anticolors save the one initiating the transition. That is, the red-quark may go to antiwhite (antilepton), antigreen, or antiblue, but not antired. In any case, baryon and lepton numbers are not conserved in these transitions. It is also curious that these transitions are not mediated by any gauge fields; it is the operator $\partial\!\!\!/_{0,6}$ itself that provides the necessary compensating charges. Recall that as color $SU(3) = Spin(6) \cap G_2$, the color symmetry $SU(3)$ is a subgroup of the (1,9)-Lorentz group, and the operator $\partial\!\!\!/_{0,6}$ is not an $SU(3)$ singlet.

An objection might be made to this on the grounds that color $SU(3)$ is an internal symmetry, and therefore ought to commute with the spacetime Clifford algebra. However, there is absolutely no evidence that $SU(3)$ must commute with any Clifford algebra larger than $\mathbf{R}_{1,3}$, that of our observable spacetime. Indeed, there is also no evidence for the extra six dimensions of $\mathbf{R}^{1,9}$. That there are an extra six dimensions, and that they carry color $SU(3)$ charges, is a prediction of model building on $\mathbf{T}$. To better understand these charges, and to figure out some conditions sufficient to rid us of the last two terms of 6.10, we must first look more closely at $SU(3)$.

6.2. More $SU(3)$.

The action of $SU(3)$ on $\rho_+ X$, where $X \in \mathbf{O}$ (or $\mathbf{C}\otimes\mathbf{O}$ or $\mathbf{T}$), was developed in section 3.6. For convenience I'll reproduce the principal results here. In particular, $SU(3)$ is generated by elements of the form: $e_{Lpq} - e_{Lrs}$, where $e_p e_q = e_r e_s$, and $p, q, r, s \neq 7$. These generators leave 1 and e_7 invariant, hence also $\Psi^0_\pm$. Their action on $\Psi^m_\mp$, m=1,2,4, can be represented as a $\mathbf{C}(3)$ action on the column $\begin{bmatrix} \Psi^1_+ \\ \Psi^2_+ \\ \Psi^4_+ \end{bmatrix}$. The corresponding matrices are:

$$\tfrac{1}{2}(e_{L34}-e_{L26}) \to \begin{bmatrix} \cdot & \cdot & \cdot \\ \cdot & \cdot & i \\ \cdot & i & \cdot \end{bmatrix}, \quad \tfrac{1}{2}(e_{L63}-e_{L24}) \to \begin{bmatrix} \cdot & \cdot & \cdot \\ \cdot & \cdot & 1 \\ \cdot & -1 & \cdot \end{bmatrix},$$

$$\tfrac{1}{2}(e_{L61}-e_{L45}) \to \begin{bmatrix} \cdot & \cdot & i \\ \cdot & \cdot & \cdot \\ i & \cdot & \cdot \end{bmatrix}, \quad \tfrac{1}{2}(e_{L56}-e_{L41}) \to \begin{bmatrix} \cdot & \cdot & 1 \\ \cdot & \cdot & \cdot \\ -1 & \cdot & \cdot \end{bmatrix},$$

$$\tfrac{1}{2}(e_{L52}-e_{L13}) \to \begin{bmatrix} \cdot & i & \cdot \\ i & \cdot & \cdot \\ \cdot & \cdot & \cdot \end{bmatrix}, \quad \tfrac{1}{2}(e_{L35}-e_{L12}) \to \begin{bmatrix} \cdot & 1 & \cdot \\ -1 & \cdot & \cdot \\ \cdot & \cdot & \cdot \end{bmatrix}, \tag{6.17}$$

$$\tfrac{1}{2}(e_{L46}-e_{L23}) \to \begin{bmatrix} \cdot & \cdot & \cdot \\ \cdot & i & \cdot \\ \cdot & \cdot & -i \end{bmatrix}, \quad \tfrac{1}{2}(e_{L23}-e_{L15}) \to \begin{bmatrix} i & \cdot & \cdot \\ \cdot & -i & \cdot \\ \cdot & \cdot & \cdot \end{bmatrix}.$$

The corresponding matrices acting on the column $\begin{bmatrix} \Psi^1_- \\ \Psi^2_- \\ \Psi^4_- \end{bmatrix}$ are the complex conjugates of those in 6.12.

$SO(6)$, hence also $SU(3)$, acts on $\mathbf{R}^{0,6}$. The action of $SU(3)$ may be determined by commuting the generators listed in 6.17 with the element

$$X = \sum_1^6 X^{p+3} i\omega e_{Lp}$$

of $\mathbf{R}_{1,9} \subseteq \mathbf{T}_L(2)$. Since these generators commute with $i\omega$, we may look only at their action on the e_{Lp}. For example, the commutator

$$[\frac{1}{2}(e_{L34}-e_{L26}), \sum_1^6 X^{p+3} e_{Lp}] = X^9 e_{L2} - X^7 e_{L3} + X^6 e_{L4} - X^5 e_{L6}. \tag{6.18}$$

Therefore, under the action 6.18,

$$\begin{bmatrix} X^4 + iX^8 \\ X^5 + iX^6 \\ X^7 + iX^9 \end{bmatrix} \to \begin{bmatrix} 0 \\ X^9 - iX^7 \\ X^6 - iX^5 \end{bmatrix} = \begin{bmatrix} \cdot & \cdot & \cdot \\ \cdot & \cdot & -i \\ \cdot & -i & \cdot \end{bmatrix} \begin{bmatrix} X^4 + iX^8 \\ X^5 + iX^6 \\ X^7 + iX^9 \end{bmatrix}. \tag{6.19}$$

Likewise the commutator

$$[\frac{1}{2}(e_{L56} - e_{L41}), \sum_{1}^{6} X^{p+3} e_{Lp}] = -X^7 e_{L1} + X^4 e_{L4} - X^9 e_{L5} + X^8 e_{L6}. \quad (6.20)$$

implies

$$\begin{bmatrix} X^4 + iX^8 \\ X^5 + iX^6 \\ X^7 + iX^9 \end{bmatrix} \to \begin{bmatrix} X^7 + iX^9 \\ 0 \\ -X^4 - iX^8 \end{bmatrix} = \begin{bmatrix} . & . & -1 \\ . & . & . \\ 1 & . & . \end{bmatrix} \begin{bmatrix} X^4 + iX^8 \\ X^5 + iX^6 \\ X^7 + iX^9 \end{bmatrix}. \quad (6.21)$$

Therefore $\begin{bmatrix} X^4 + iX^8 \\ X^5 + iX^6 \\ X^7 + iX^9 \end{bmatrix}$ transforms with respect to $SU(3)$ under a representation the inverse of that in table 6.17, which acts on $\rho_+ \Psi \rho_-$, a quark-matter spinor. Likewise, $\begin{bmatrix} X^4 - iX^8 \\ X^5 - iX^6 \\ X^7 - iX^9 \end{bmatrix}$ transforms under a representation the inverse of that on $\rho_- \Psi \rho_+$, an antiquark-antimatter spinor.

6.3. Freedom from Matter-Antimatter Mixing.

If f is a suitably smooth function of a complex variable, $z = x + iy$, then

$$(\partial_x + i\partial_y) f(x + iy) = 0. \quad (6.22)$$

Likewise,

$$(\partial_x - i\partial_y) f(x - iy) = 0. \quad (6.23)$$

The parts of $\mathcal{L}_{1,9}$ (see 6.5) that give rise to matter-antimatter mixing are those in which the differential operator $\not\partial_{0,6}$ appears. In 6.9 it was shown that $\not\partial_{0,6}(\rho_+ \Psi)$ can be rewritten using the operators $\partial^4 \pm i\partial^8$, $\partial^5 \pm i\partial^6$, and $\partial^7 \pm i\partial^9$. Combining 6.14 with 6.22 and 6.23, it is apparent that if

$$\boxed{\begin{aligned} \Psi_+^0 &= \Psi_+^0(x^\mu, x^4 + ix^8, x^5 + ix^6, x^7 + ix^9), \\ \Psi_+^1 &= \Psi_+^1(x^\mu, x^4 + ix^8, x^5 - ix^6, x^7 - ix^9), \\ \Psi_+^2 &= \Psi_+^2(x^\mu, x^4 - ix^8, x^5 + ix^6, x^7 - ix^9), \\ \Psi_+^4 &= \Psi_+^4(x^\mu, x^4 - ix^8, x^5 - ix^6, x^7 + ix^9), \end{aligned}} \quad (6.24)$$

where $\mu = 0,1,2,3$, then

$$\partial\!\!\!/_{0,6}(\rho_+\Psi) = 0 \tag{6.25}$$

identically (for the corresponding bits of $\rho_-\Psi$ we would require the complex conjugate dependencies).

The group $U(3)$ acts transitively on S^5, the 5-sphere determined by

$$\sum_4^9 (x^p)^2 = r^2.$$

Consequently, since Ψ_+^0 is supposed to be $U(3)$ invariant, we should expect that its only dependence on x^p, $p = 4, ..., 9$, be radial (a function of r). But this is not possible given the functional dependence in 6.24, for no nontrivial function of the complex triple

$$(x^4 + ix^8, x^5 + ix^6, x^7 + ix^9) = (z^1, z^2, z^4)$$

can be dependent only on

$$r = (z^1 z^{1*} + z^2 z^{2*} + z^4 z^{4*})^{\frac{1}{2}}.$$

Arguing along these phenomenological lines we might then expect Ψ_+^0 to be independent entirely of x^p, $p = 4, ..., 9$.

Likewise we might expect Ψ_+^1, identified as the red quark, to be independent of x^5, x^6, x^7, x^9, which transform nontrivially with respect to the $SU(2)$ subgroup of $SU(3)$ that leaves red (ie., e_1, e_5, and $x^1 + ix^5$) invariant. Carrying on like this leads to reduced dependencies:

$$\boxed{\begin{aligned} \Psi_+^0 &= \Psi_+^0(x^\mu, ., ., .), \\ \Psi_+^1 &= \Psi_+^1(x^\mu, x^4 + ix^8, ., .), \\ \Psi_+^2 &= \Psi_+^2(x^\mu, ., x^5 + ix^6, .), \\ \Psi_+^4 &= \Psi_+^4(x^\mu, ., ., x^7 + ix^9), \end{aligned}} \tag{6.26}$$

where , ., indicates independence of that coordinate.

Because of the results of section 6.2 it is clear that the $\mathbf{R}_{0,6}$ dependencies in 6.24 have complicated $SU(3)$ transformation properties. The dependencies 6.26 are phenomenologically preferable. Let $\Psi_+^{quark} = \begin{bmatrix} \Psi_+^1 \\ \Psi_+^2 \\ \Psi_+^4 \end{bmatrix}$, and

$Z = \begin{bmatrix} X^4 + iX^8 \\ X^5 + iX^6 \\ X^7 + iX^9 \end{bmatrix}$. Then if $U \in SU(3)$, we have the transformation

$$\Psi_+^{quark}(x^\mu, Z) \to U\Psi_+^{quark}(x^\mu, U^{-1}Z), \tag{6.27}$$

where even after such mixing, for example, we expect $(U\Psi_+^{quark})^1$ (first component) to be independent of the second two components of $U^{-1}Z$.

Finally, the dependencies 6.26 (or 6.24) reduce $\mathcal{L}_{1,9}$ to

$$\boxed{\mathcal{L}_{1,9} =< \rho_+\Psi, \not{\partial}_{1,3}\rho_+\Psi > + < \rho_-\Psi, \not{\partial}_{1,3}\rho_-\Psi >,} \tag{6.28}$$

which is where we want to be. Although I am not a string theorist, it is interesting that in both string theory and the context developed here the space $\mathbf{R}^{1,9}$ should play a prominent role. Morevover, in string theory the extra six dimensions are balled up into a small complex 3-manifold, inorder to render them unobservable, and in the $\mathbf{T}$ context, inorder to be rid of matter-antimatter mixing, the extra six dimensions should also appear as a complex triple (although it is unclear if there is a unique compactification arising from this context, much less if there is an associated $SU(3)$ holonomy group).

6.4. (1,9)-Scalar Lagrangian.

Redefine

$$\boxed{\mathcal{L}_\phi = \textstyle\sum_{a=0}^{9} < \partial_a\phi, \partial^a\phi > -\mu^2 < \phi, \phi > -\lambda < \phi, \phi >^2 .} \tag{6.29}$$

This is the obvious generalization of 5.1 from $\mathbf{R}^{1,3}$ to $\mathbf{R}^{1,9}$. That is, now it is assumed

$$\phi = \phi(x^a), a = 0, 1, ..., 9, \tag{6.30}$$

rather than merely $\phi(x^\mu), \mu = 0, 1, 2, 3$. Euler's equations,

$$\frac{\delta\mathcal{L}}{\delta\phi_i} = \sum_{a=0}^{9} \frac{\partial}{\partial x^a}\left[\frac{\delta\mathcal{L}}{\delta(\partial\phi_i/\partial x^a)}\right], \tag{6.31}$$

applied to 6.24, imply

$$-2\mu^2\phi_i - 4\lambda\phi_i < \phi, \phi >= 2\sum_0^9(\partial_a\partial^a)\phi_i.$$

Therefore

$$\sum_{0}^{3}(\partial_\mu \partial^\mu)\phi_i + \sum_{4}^{9}(\partial_p \partial^p)\phi_i + [\mu^2 + 2\lambda < \phi, \phi >]\phi_i = 0. \tag{6.32}$$

This is only unconventional in the second term. Note however that if f is a function of $x \pm iy$, with $\partial_x^2 f$ and $\partial_y^2 f$ defined and continuous, then

$$(\partial_x^2 + \partial_y^2) f(x \pm iy) = 0. \tag{6.33}$$

Therefore, if ϕ is, for example, dependent on x^p, $p = 4, ..., 9$ as Ψ_+^0 in 6.24, then

$$\sum_{4}^{9}(\partial_p \partial^p)\phi_i = 0 \tag{6.34}$$

identically, and 6.32 reduces to the conventional form (obviously there are other dependencies on the extra six dimensions that would do as well). $\mathcal{L}_\phi$, however, does not. If we wish it to, then we need take the same extra steps taken in the last section to get from 6.24 to 6.26. That is, as ϕ is an $SU(3)$ scalar, we might expect it to be independent of the $x^p, p = 4, ..., 9$, which carry $SU(3)$ charges. In which case, of course, 6.29 reduces to 5.1, and we are back where we started from, quite possibly where we'd like to be. As Higgs scalar fields are presently unobserved, much that could be said about them must be considered conjectural.

6.5. Charge Conjugation on $\mathbf{T}_L(2)$.

The Clifford algebra $\mathbf{R}_{1,9}$, which is represented in 4.4 by elements of $\mathbf{T}_L(2)$, is also representable by $\mathbf{R}(32)$, to which it is isomorphic. The spinors of this latter representation are 32×1 real columns, the so-called Majorana spinors, and $\mathbf{R}_{1,9}$ maps this space to itself. The spinors of $\mathbf{T}_L(2)$ are elements of the 128-dimensional $\mathbf{T}^2$, the space of 2×1 columns over 64-dimensional $\mathbf{T}$. This space has 4 times the dimensionality of an $\mathbf{R}_{1,9}$ Majorana spinor.

Half of that extra dimensionality arises from the quaternions, since $\mathbf{Q}_R$ is distinct from and commutes with $\mathbf{T}_L(2)$. In chapter 3 the idempotents

$$\lambda_\pm = \frac{1}{2}(1 \pm i(x^2 q_2 + x^3 i e_7 q_3)) \tag{6.35}$$

were used in conjunction with $\rho_\pm = \frac{1}{2}(1 \pm ie_7)$ to project the lepton, quark, antilepton and antiquark parts of $\mathbf{T}^2$ spinors. In particular, if

$$\Psi \in \mathbf{T}^2, \tag{6.36}$$

then $\Psi\lambda_\pm$ are 64-dimensional $\mathbf{R}_{1,9}$ Dirac spinors, the complexification of Majorana spinors (and an isodoublet; see chapters 3 and 4).

The complexification of $\mathbf{R}(32)$ is $\mathbf{C}(32)$, the (1,9)-spacetime Dirac algebra, with 32×1 complex column spinors. Add such a spinor to its complex conjugate and you end up with a 32×1 real spinor, which is Majorana if $\mathbf{R}_{1,9}$ is represented by real matrices. Take an element of $\mathbf{C}(32)$ and add it to its complex conjugate and you get an element of $\mathbf{R}(32)\simeq \mathbf{R}_{1,9}$, which of course is complex conjugate invariant. In this case complex conjugation and charge conjugation, which by definition leaves Majorana spinors invariant, are the same operation.

This simple definition of charge conjugation can be used to construct the charge conjugation map on $\mathbf{T}_L(2)$ and its spinors. In particular we require first an automorphism on $\mathbf{T}_L(2)$ that sends

$$i_L \to -i_L,$$

but that leaves the $\mathbf{R}_{1,9}$ subalgebra of $\mathbf{T}_L(2)$, given in 4.4, invariant. Every element of $\mathbf{T}_L(2)$ can be written as $M + i_L N$, where $M, N \in \mathbf{R}_{1,9}$. Define

$$\boxed{(M + i_L N)^\# = M - i_L N.} \tag{6.37}$$

The $\mathbf{T}_L(2)$ automorphism 6.36 can be extended to the spinor space $\mathbf{T}^2$, but not uniquely. As a first step, note that since

$$\epsilon = \begin{bmatrix} 1 & 0 \\ 0 & 1 \end{bmatrix}, \alpha = \begin{bmatrix} 1 & 0 \\ 0 & -1 \end{bmatrix}, \beta = \begin{bmatrix} 0 & 1 \\ 1 & 0 \end{bmatrix}, \omega = \begin{bmatrix} 0 & 1 \\ -1 & 0 \end{bmatrix} \in \mathbf{R}_{1,9},$$

it is therefore true that

$$\epsilon^{\#} = \epsilon,\ \alpha^{\#} = \alpha,\ \beta^{\#} = \beta,\ \omega^{\#} = \omega. \tag{6.38}$$

So if $A, B, C, D \in \mathbf{T}_L$, then

$$\begin{bmatrix} A & C \\ D & B \end{bmatrix}^{\#} = \begin{bmatrix} A^{\#} & C^{\#} \\ D^{\#} & B^{\#} \end{bmatrix}. \tag{6.39}$$

That is, the action of this conjugation on $\mathbf{T}_L$ fully determines its action on all of $\mathbf{T}_L(2)$.

An examination of the Clifford algebra 4.4, together with $i_L^{\#} = -i_L$, determines the following actions on $\mathbf{T}_L$:

$$\begin{aligned}
i_L^{\#} &= -i_L, \\
q_{Lj}^{\#} &= +q_{Lj}, \\
e_{Lp}^{\#} &= -e_{Lp}, && p \in \{1, ..., 6\}, \\
e_{L7}^{\#} &= +e_{L7}. \\
e_{Lpq}^{\#} &= +e_{Lpq}, && p, q \in \{1, ..., 6\}, \\
e_{Lp7}^{\#} &= -e_{Lp7}, && p \in \{1, ..., 6\}, \\
e_{Lpq7}^{\#} &= +e_{Lpq7}, && p, q \in \{1, ..., 6\}, \\
e_{Lpqr}^{\#} &= -e_{Lpqr}, && p, q, r \in \{1, ..., 6\},
\end{aligned} \tag{6.40}$$

Together with products of these elements, this completely determines the $A \rightarrow A^{\#}$ conjugation on $\mathbf{T}_L$, hence also on $\mathbf{T}_L(2)$.

6.6. Charge Conjugation on $\mathbf{T}^2$.

If $x, y \in \mathbf{T}$, then we expect the extension of the Dirac algebra conjugation 6.39,40 to the corresponding Dirac spinors to act as follows:

$$\begin{bmatrix} x \\ y \end{bmatrix}^{\#} = \begin{bmatrix} x^{\#} \\ y^{\#} \end{bmatrix}. \tag{6.41}$$

This is a consequence of 6.39. We also expect

$$\begin{bmatrix} x \\ y \end{bmatrix}^{\#} = \left(\begin{bmatrix} x_L & 0 \\ 0 & y_L \end{bmatrix} \begin{bmatrix} 1 \\ 1 \end{bmatrix} \right)^{\#} = \begin{bmatrix} x_L^{\#} & 0 \\ 0 & y_L^{\#} \end{bmatrix} \begin{bmatrix} 1^{\#} \\ 1^{\#} \end{bmatrix}. \tag{6.42}$$

Therefore we may concentrate on just $\mathbf{T}_L$ and $\mathbf{T}$, the action of the conjugation on the former determined by 6.40.

[Note: while we expect $1_L^{\#} = 1_L$, we will find that $1^{\#} \neq 1$, where it should be remembered that $1 \in \mathbf{T}$ is an element of the spinor space.]

We begin with the action on $x \in \mathbf{T}$,

$$x^{\#} = (x_L[1])^{\#} = x_L^{\#}[1^{\#}]. \tag{6.43}$$

In particular observe that

$$e_6^{\#} = e_{L6}^{\#}[1^{\#}] = -e_{L6}[1^{\#}],$$

and also that

$$e_6^{\#} = (e_1 e_2)^{\#} = (e_{L12}[1])^{\#} = e_{L12}^{\#}[1^{\#}] = +e_{L12}[1^{\#}].$$

In general, if $p, q, r \in \{1, ..., 6\}$ are distinct, $e_p e_q = e_r$, then

$$e_r^{\#} = -e_{Lr}[1^{\#}] = +e_{Lpq}[1^{\#}]. \tag{6.44}$$

This necessarily implies that

$$\boxed{1^{\#} = k e_7,} \tag{6.45}$$

where k is an as yet undetermined element of $\mathbf{C} \otimes \mathbf{Q}$.

Since $x \to x^{\#}$ is supposed to be a kind of complex conjugation on the spinor x, we expect

$$(x^{\#})^{\#} = x. \tag{6.46}$$

Therefore,

$$1 = (1^{\#})^{\#} = (ke_7)^{\#} = k_L^{\#} e_{L7}^{\#}[1^{\#}] = k_L^{\#} e_{L7}[ke_7] = -k_L^{\#}[k]. \tag{6.47}$$

So

$$k_L^{\#}[k] = -1. \tag{6.48}$$

Using 6.40 we determine

$$\begin{aligned} 1_L^{\#}[1] &= 1_L[1] = i_L^{\#}[i] = -i_L[i] = 1, \\ q_{Lj}^{\#}[q_j] &= q_{Lj}[q_j] = (i_L q_{Lj})^{\#}[iq_j] = -i_L q_{Lj}[iq_j] = -1. \end{aligned} \tag{6.49}$$

From 6.48,49 it is not difficult to determine that

$$k = e^{i\theta}\vec{u}, \tag{6.50}$$

where $\vec{u}$ is a unit quaternion. For the sake of simplicity we will set

$$k = q_2, \tag{6.51}$$

so that in general, the charge conjugate of $x \in \mathbf{T}$ is

$$\boxed{x^{\#} = x_L^{\#}[q_2 e_7] = q_2 e_7 M_3(x),} \tag{6.52}$$

where $M_3(x) = -q_2 x^* q_2$ was defined in chapter 3, and in chapter 4 we saw that it reverses all charges (the last equality above is left as an exercise for the reader).

[Note: Although we are fixing k, it is interesting that given the form 6.50, $k \in U(1) \times SU(2)$, in which case 6.45 has a $U(1) \times SU(2) \times SU(3)$ form invariance, $SU(3) \subset G_2$ the stability group of e_7.]

Therefore, for example, if

$$\begin{aligned} x &= \tfrac{1}{2}(x\lambda_\pm + q_2 e_7 M_3(x\lambda_\pm)), \\ y &= \tfrac{1}{2}(y\lambda_\pm + q_2 e_7 M_3(y\lambda_\pm)), \end{aligned} \tag{6.53}$$

then $\begin{bmatrix} x \\ y \end{bmatrix}$ is a Majorana spinor, and $\begin{bmatrix} x \\ 0 \end{bmatrix}$ and $\begin{bmatrix} 0 \\ y \end{bmatrix}$ are Majorana-Weyl, $\mathbf{R}_{1,9}$ being one of that class of Clifford algebras the spinors of which can be both Majorana and Weyl.

As an example of the action of the charge conjugation 6.52, recall that if $\Psi \in \mathbf{T}^2$ is a general complex (1,9)-spinor field, then

$$\rho_+\Psi\rho_-\lambda_+ : \; I_3 = \frac{1}{2},\, Y_2 = \frac{1}{2},\, Y_3 = -\frac{1}{3},\, Q = \frac{2}{3} \tag{6.54}$$

is a charge $\frac{2}{3}$ quark color triplet spinor field (see 4.17). In this case

$$\begin{aligned}(\rho_+\Psi\rho_-\lambda_+)^\# &= q_2 e_7 M_3(\rho_+\Psi\rho_-\lambda_+) \\ &= iq_2\rho_- M_3(\Psi)\rho_+\lambda_-,\end{aligned} \tag{6.55}$$

where

$$\rho_-\Psi'\rho_+\lambda_- : \; I_3 = -\frac{1}{2},\, Y_2 = -\frac{1}{2},\, Y_3 = \frac{1}{3},\, Q = -\frac{2}{3} \tag{6.56}$$

is a charge $-\frac{2}{3}$ antiquark.

THE MEANING OF MAJORANA

The general field Ψ may be written

$$\boxed{\Psi = \rho_+\Psi + \rho_-\Psi = F_1 + F_2^\#,} \tag{6.57}$$

where F_1 and F_2 are complete leptoquark families, and $F_2^\#$ is the antifamily of F_2:

$$F_2^\# = q_2 e_7 M_3(F_2) = iq_2 M_3(F_2). \tag{6.58}$$

We can now see the difference between a (1,9)-Dirac spinor, and a (1,9)-Majorana spinor. The former, $F_1 + F_2^\#$ is actually an $SU(2)$ doublet of (1,9)-Dirac spinors, and its family and antifamily parts are independent. The latter, from 6.53, is

$$\begin{aligned}\tfrac{1}{2}(\Psi + \Psi^\#) &= \tfrac{1}{2}(F_1 + F_2^\# + (F_1 + F_2^\#)^\#) \\ &= \tfrac{1}{2}(F_1 + F_2 + (F_1 + F_2)^\#),\end{aligned} \tag{6.59}$$

where now we see that the matter and antimatter halves are conjugates of each other. The (1,9)-Dirac spinors allow for weak mixing a la section 4.5, and presumably contain information re family replication. A (1,9)-Majorana spinor still consists of a family of (1,3)-Dirac spinors (NOT (1,3)-Majorana!), summed with its charged conjugate, but offers no opening for weak mixing, nor any hints as to family replication. I emphasize these distinctions as in the context of super string theory only (1,9)-Majorana spinors are employed. In the next section we will look a little closer at this context.

6.7. 10 Other Dimensions.

10-dimensional space is by-and-large an ineluctable aspect of the application of **T**. It is a part of the message of the mathematics, and coincidentally it also plays a key role in string theory, the largest contemporary arena of theoretical physics.

In classical string theory there are four critical dimensions: 3, 4, 6, and 10. These are naturally linked to **R**, **C**, **Q**, and **O**. To see this note that

$$det\begin{bmatrix} x^0+x^3 & x^1+ix^2 \\ x^1-ix^2 & x^0-x^3 \end{bmatrix} = (x^0)^2-(x^1)^2-(x^2)^2-(x^3)^2. \qquad (6.60)$$

This link of the Hermitian matrices of $\mathbf{C}(2)$ to (1,3)-space-time is well-known, and it establishes the paradigm for linking the Hermitian matrices of $\mathbf{R}(2)$, $\mathbf{Q}(2)$, and $\mathbf{O}(2)$ to space-times of other dimensions. In particular,

$$det\begin{bmatrix} x^0+x^2 & x^1 \\ x^1 & x^0-x^2 \end{bmatrix} = (x^0)^2-(x^1)^2-(x^2)^2 \qquad (6.61)$$

links $\mathbf{R}(2)$ to $\mathbf{R}^{1,2}$;

$$det\begin{bmatrix} x^0+x^5 & x^4+\sum_1^3 x^i q_i \\ x^4-\sum_1^3 x^i q_i & x^0-x^5 \end{bmatrix} \qquad (6.62)$$
$$= (x^0)^2-(x^1)^2-(x^2)^2-(x^3)^2-(x^4)^2-(x^5)^2$$

links $\mathbf{Q}(2)$ to $\mathbf{R}^{1,5}$; and

$$det\begin{bmatrix} x^0+x^9 & x^8+\sum_1^7 x^a e_a \\ x^8-\sum_1^7 x^a e_a & x^0-x^9 \end{bmatrix}$$
$$= (x^0)^2-(x^1)^2-(x^2)^2-(x^3)^2-(x^4)^2 \qquad (6.63)$$
$$-(x^5)^2-(x^6)^2-(x^7)^2-(x^8)^2-(x^9)^2$$

links $\mathbf{O}(2)$ to $\mathbf{R}^{1,9}$.

These links of the 2×2 Hermitian matrices over **R**, **C**, **Q**, and **O** to 3, 4, 6, and 10-dimensional spacetimes allow an easy algebraic construction of twistors (conformal group spinors) for these spacetimes. In particular, let

$$\eta,\ \xi \in \mathbf{K}_m, \qquad (6.64)$$

where

$$\mathbf{K}_0 = \mathbf{R},\ \mathbf{K}_1 = \mathbf{C},\ \mathbf{K}_2 = \mathbf{Q},\ \mathbf{K}_3 = \mathbf{O}.$$

If we define

$$\pi = \begin{bmatrix} \eta \\ \xi \end{bmatrix}, \tag{6.65}$$

then

$$\pi\pi^\dagger = \begin{bmatrix} \eta\eta^\dagger & \eta\xi^\dagger \\ \xi\eta^\dagger & \xi\xi^\dagger \end{bmatrix} \tag{6.66}$$

is a 2×2 Hermitian matrix over $\mathbf{K}_m$, and

$$det(\pi\pi^\dagger) = 0,$$

so $\pi\pi^\dagger$ represents a $(1, 1 + 2^m)$-spacetime vector in the forward light-cone (forward because the time component,

$$\eta\eta^\dagger + \xi\xi^\dagger \geq 0).$$

Lorentz transformations on the spacetime vector can be achieved via an $SL(2, \mathbf{K}_m)$ action on π **[14]**. In particular, define $sl(2, \mathbf{K}_m)$, the Lie algebra of $SL(2, \mathbf{K}_m)$, to be the L-commutator closure of the set of traceless 2×2 matrices over $\mathbf{K}_m$. By L-commutator closure I mean that once the traceless 2×2 matrices over $\mathbf{K}_m$ are formed, we map the components to $\mathbf{K}_{Lm}$, the left adjoint algebra (only in the case of **O** will this make a difference), then use commutation to find the elements needed to close the algebra. Bases for $sl(2, \mathbf{K}_m)$, $m = 1, 2, 3$, are given below:

$$sl(2,\mathbf{C}): \quad \begin{bmatrix} 1_L & 0 \\ 0 & -1_L \end{bmatrix}, \begin{bmatrix} i_L & 0 \\ 0 & -i_L \end{bmatrix},$$

$$\begin{bmatrix} 0 & 1_L \\ 1_L & 0 \end{bmatrix}, \quad \begin{bmatrix} 0 & i_L \\ i_L & 0 \end{bmatrix}, \tag{6.67}$$

$$\begin{bmatrix} 0 & 1_L \\ -1_L & 0 \end{bmatrix}, \begin{bmatrix} 0 & i_L \\ -i_L & 0 \end{bmatrix};$$

$$sl(2,\mathbf{Q}): \quad \begin{bmatrix} 1_L & 0 \\ 0 & -1_L \end{bmatrix}, \begin{bmatrix} q_{Lj} & 0 \\ 0 & -q_{Lj} \end{bmatrix},$$

$$\begin{bmatrix} 0 & 1_L \\ 1_L & 0 \end{bmatrix}, \quad \begin{bmatrix} 0 & q_{Lj} \\ q_{Lj} & 0 \end{bmatrix},$$

$$\begin{bmatrix} 0 & 1_L \\ -1_L & 0 \end{bmatrix}, \begin{bmatrix} 0 & q_{Lj} \\ -q_{Lj} & 0 \end{bmatrix}, \tag{6.68}$$

$$\Longrightarrow \qquad \begin{bmatrix} q_{Lj} & 0 \\ 0 & q_{Lj} \end{bmatrix};$$

$$sl(2,\mathbf{O}): \quad \begin{bmatrix} 1_L & 0 \\ 0 & -1_L \end{bmatrix}, \quad \begin{bmatrix} e_{La} & 0 \\ 0 & -e_{La} \end{bmatrix},$$

$$\begin{bmatrix} 0 & 1_L \\ 1_L & 0 \end{bmatrix}, \quad \begin{bmatrix} 0 & e_{La} \\ e_{La} & 0 \end{bmatrix},$$

$$\begin{bmatrix} 0 & 1_L \\ -1_L & 0 \end{bmatrix}, \begin{bmatrix} 0 & e_{La} \\ -e_{La} & 0 \end{bmatrix}, \tag{6.69}$$

$$\Longrightarrow \qquad \begin{bmatrix} e_{Lab} & 0 \\ 0 & e_{Lab} \end{bmatrix}.$$

In each of these cases there are respectively

$$3(2^m) = 6,\ 12,\ 24$$

traceless matrices over $\mathbf{K}_{Lm}$.

$$\boxed{sl(2,\mathbf{C}) \simeq so(1,3)}$$

closes on its 6 traceless generators. But because $\mathbf{Q}$ is noncommutative,

$$\boxed{sl(2,\mathbf{Q}) \simeq so(1,5)}$$

requires 3 additional generators, making $3 + 12 = 15$ in all. And because $\mathbf{O}$ is both noncommutative and nonassociative,

$$\boxed{sl(2,\mathbf{O}) \simeq so(1,9)}$$

requires another 21 generators, making $24 + 21 = 45$ in all.

The Lie groups $SL(2,\mathbf{K}_m)$ generated by the $sl(2,\mathbf{K}_m)$ above act on the π given in 6.65, and this results in a Lorentz transformation of the $(1, 1+2^m)$-spacetime vector $\pi\pi^\dagger$.

π is not yet a conformal group spinor. In particular, consider the case of $\mathbf{K}_3 = \mathbf{O}$. Define in $\mathbf{O}_L(2)$,

$$\begin{aligned} \sigma_0 &= \begin{bmatrix} 1_L & 0 \\ 0 & 1_L \end{bmatrix}, \quad \sigma_a = \begin{bmatrix} 0 & e_{La} \\ -e_{La} & 0 \end{bmatrix}, \quad a = 1, ..., 7, \\ \sigma_8 &= \begin{bmatrix} 0 & 1_L \\ 1_L & 0 \end{bmatrix}, \quad \sigma_9 = \begin{bmatrix} 1_L & 0 \\ 0 & -1_L \end{bmatrix}. \end{aligned} \tag{6.70}$$

Then

$$\pi\pi^\dagger = p = \sum_0^9 p^k \sigma_k \tag{6.71}$$

is a (1,9)-momentum vector for a massless particle.

With the σ_k we can construct a representation of $so(2,10)$, the conformal group of $\mathbf{R}^{1,9}$. Let

$$\sigma_{rs} = \frac{1}{2}[\sigma_r, \sigma_s],\ r, s \in \{1, ..., 9\}.$$

Then the 66 distinct matrices

$$\begin{bmatrix} \sigma_{rs} & 0 \\ 0 & \sigma_{rs} \end{bmatrix}, \begin{bmatrix} \sigma_k & 0 \\ 0 & -\sigma_k \end{bmatrix}, \begin{bmatrix} 0 & \sigma_k \\ 0 & 0 \end{bmatrix}, \begin{bmatrix} 0 & 0 \\ \sigma_k & 0 \end{bmatrix}, \tag{6.72}$$

($k = 0, ..., 9$) represent $so(2, 10)$. Let

$$X = \sum_0^9 X^k \sigma_k, \tag{6.73}$$

then

$$exp \begin{bmatrix} 0 & \\ X & 0 \end{bmatrix} = \begin{bmatrix} 1 & 0 \\ X & 1 \end{bmatrix} \tag{6.74}$$

can be chosen to represent a translation. In this case,

$$\begin{bmatrix} 1 & 0 \\ X & 1 \end{bmatrix} \begin{bmatrix} \pi \\ 0 \end{bmatrix} = \begin{bmatrix} \pi \\ X\pi \end{bmatrix}, \tag{6.75}$$

where the full algebraic twistor $\begin{bmatrix} \pi \\ X\pi \end{bmatrix}$ is a conformal group spinor.

Similar constructions are possible for the other division algebras. Octonionic twistors have been exploited by [8].

THE CLIFFORD ALGEBRA

More generally, in octonionic approaches to string theory over $\mathbf{R}^{1,9}$, an

$$\mathbf{O}_L(4) \simeq \mathbf{R}_{1,9} \tag{6.76}$$

representation of the (1,9)-Clifford algebra is used.

[Note: In [7] and [8] things looks different. They start with $\mathbf{O}(4)$ and nest matrices, whereas I nest $\mathbf{O}$ to get $\mathbf{O}_L$, then incorporate $\mathbf{O}_L$ into matrices, which has the advantage of being generally applicable.]

Given the σ_k, $k = 0, ..., 9$, in 6.70, we can define $\mathbf{R}_{1,9}$ γ-matrices as follows:

$$\gamma_k = \begin{bmatrix} 0 & \sigma_k \\ \sigma_k^- & 0 \end{bmatrix}, \tag{6.77}$$

where as usual,

$$\sigma_0^- = \sigma_0, \ \ \sigma_r^- = -\sigma_r, \ r = 1, ..., 9.$$

The spinors of the resultant Clifford algebra are elements of $\mathbf{O}^4$ (4×1 columns over $\mathbf{O}$). These $(4)(8) = 32$-dimensional spinors are (1,9)-Majorana spinors. And as has been pointed out more than once, in this case Weyl spinors may be projected without complexification, as the (1,9)-pseudoscalar is

$$\gamma_0\gamma_1...\gamma_9 = \begin{bmatrix} \sigma_0 & 0 \\ 0 & -\sigma_0 \end{bmatrix} \tag{6.78}$$

(to within a sign), from which the chiral projectors,

$$\begin{bmatrix} \sigma_0 & 0 \\ 0 & 0 \end{bmatrix}, \ \begin{bmatrix} 0 & 0 \\ 0 & \sigma_0 \end{bmatrix}, \tag{6.79}$$

are easily contructable.

This context differs from that developed in this monograph in another important sense. Although six of the γ_k defined in 6.75 are associated with the octonion directions e_p, $p = 1, ..., 6$, this fact is not linked to color $SU(3)$. Although these ten γ-matrices can be reduced via the projectors $\rho_{L\pm} = \frac{1}{2}(1 \pm ie_{L7})$ to $\mathbf{R}_{1,3}$ γ-matrices, just as was done in the $\mathbf{T}$ context, the reality condition required of the string context evidently precludes this form of dimensional reduction. This reduction is so neat and pretty, and links so nicely to the color $SU(3)$ leptoquark multiplet structure, it is hard for me to believe that it can not play a role in the elegant octonionic constructions of string theory.

There are of course other ways to distinguish the octonionic string context, with $\mathbf{O}_L(4) \simeq \mathbf{R}_{1,9}$ and $\mathbf{O}^4$ spinors, from my context, with $\mathbf{T}_L(2) \simeq \mathbf{C} \otimes \mathbf{R}_{1,9}$, $\mathbf{T}^2$ spinors, and internal $\mathbf{Q}_R$. While it is true that

$$\mathbf{O}_L(4) \subset \mathbf{T}_L(2),$$

as pointed out in the last section, the Majorana spinors of $\mathbf{O}_L(4)$ do not have as rich a family structure as the Dirac spinors of $\mathbf{T}_L(2)$, and of course the $\mathbf{O}_L(4)$ spinors are $SU(2)$ isosinglets, because there is no $\mathbf{Q}_R$ to generate the $SU(2)$ internal symmetry. In string theory the standard symmetry is generally assumed to be a subgroup of some larger GUT group, like E_8.

This curious connection of the critical dimensions of string theory to the division algebras has spawned a minor industry devoted to the search for deeper meaning. The applications [7] [8] simplify, generalize, and explicate some of the results of classical 10-dimensional string theory. Given their results it seems clear that there is a connection of the octonions to string theory - leaving open the ultimate connection of string theory to reality. In pursuing such connections we must eventually address the question of what it means for an algebra like **O** to be "applied" to physics? Which is fundamental, the underlying mathematics, or the physics? If **O** merely provides a simplifying representation, the universe takes on a prosaic aspect that to me is disappointing. If on the other hand the design of reality arises from a web of select and beautiful mathematical notions (a poor word, but there may not be one that is more appropriate), a part of which is division algebra theory, then the universe, in my possibly perverse view, acquires elegance, even refulgence. Perhaps physicists resist this idea as it would pull their science out of the realm of the cause and effect of things on each other - the what-happens-if-this-happens view of physical truth - and into a realm from the perspective of which almost all that they do is secondary to something more fundamental. Physicists are prone to smugness in pointing out that chemistry is a less fundamental science, derivable from physics. I would suggest we are on the doorstep of a new science, one more fundamental than physics as it is conventionally defined, and from which physics may be derived. Physics is historically a nuts-and-bolts science, but Truth is not. Truth explains the nuts and bolts, and not in terms of other nuts and bolts.

7. Doorways.

7.1. Moufang and other Identities.

We have defined

$$\begin{aligned}
P_0 &= \tfrac{1}{8}(1 + e_{L476} + e_{L517} + e_{L621} + e_{L732} + e_{L143} + e_{L254} + e_{L365}),\\
P_1 &= \tfrac{1}{8}(1 - e_{L476} + e_{L517} + e_{L621} - e_{L732} + e_{L143} - e_{L254} - e_{L365}),\\
P_2 &= \tfrac{1}{8}(1 - e_{L476} - e_{L517} + e_{L621} + e_{L732} - e_{L143} + e_{L254} - e_{L365}),\\
P_3 &= \tfrac{1}{8}(1 - e_{L476} - e_{L517} - e_{L621} + e_{L732} + e_{L143} - e_{L254} + e_{L365}),\\
P_4 &= \tfrac{1}{8}(1 + e_{L476} - e_{L517} - e_{L621} - e_{L732} + e_{L143} + e_{L254} - e_{L365}),\\
P_5 &= \tfrac{1}{8}(1 - e_{L476} + e_{L517} - e_{L621} - e_{L732} - e_{L143} + e_{L254} + e_{L365}),\\
P_6 &= \tfrac{1}{8}(1 + e_{L476} - e_{L517} + e_{L621} - e_{L732} - e_{L143} - e_{L254} + e_{L365}),\\
P_7 &= \tfrac{1}{8}(1 + e_{L476} + e_{L517} - e_{L621} + e_{L732} - e_{L143} - e_{L254} - e_{L365}).
\end{aligned} \tag{7.1}$$

(We will dispense with the subscript L on the identity 1_L of $\mathbf{O}_L$ in most of what follows.) These satisfy

$$\sum_{a=0}^{7} P_a = 1, \tag{7.2}$$

and

$$P_a P_b = \delta_{ab} P_b. \tag{7.3}$$

They are orthogonal primitive idempotents resolving the identity of $\mathbf{O}_L$. They are related as follows ($a \in \{0, 1, ..., 7\}$):

$$P_a = e_{La} P_0 e_{La}^\dagger; \tag{7.4}$$

$$P_a e_{La} = e_{La} P_0; \tag{7.5}$$

$$e_{La} P_a = P_0 e_{La}; \tag{7.6}$$

if $e_a e_b = e_c$ $(a, b, c \in \{1, ..., 7\})$, then

$$e_{La} P_0 e_{Lb} = -e_{Lb} P_c e_{La}. \tag{7.7}$$

In what follows, let $x, y, z \in \mathbf{O}$, so that

$$\begin{array}{lll} x = \sum_0^7 x^a e_a, & y = \sum_0^7 y^a e_a, & z = \sum_0^7 z^a e_a, \\ x_L = \sum_0^7 x^a e_{La}, & y_L = \sum_0^7 y^a e_{La}, & z_L = \sum_0^7 z^a e_{La}, \\ x_R = \sum_0^7 x^a e_{Ra}, & y_R = \sum_0^7 y^a e_{Ra}, & z_R = \sum_0^7 z^a e_{Ra}. \end{array} \tag{7.8}$$

For $a = 0, ..., 7$,

$$\boxed{P_a[x] = x^a e_a \text{ (no sum).}} \tag{7.9}$$

In particular,

$$\boxed{P_0[x] = x^0,} \tag{7.10}$$

the real part of x.

Define

$$T_a = 2P_a - 1. \tag{7.11}$$

Then

$$\boxed{\begin{aligned} T_0[x] &= 2x^0 - x \\ &= x^0 - x^1 e_1 - ... - x^7 e_7 \\ &= x^\dagger. \end{aligned}} \tag{7.12}$$

So

$$T_0[xy] = (xy)^\dagger = y^\dagger x^\dagger = T_0[y] T_0[x]. \tag{7.13}$$

Therefore,

$$T_0 x_L[y] = T_0[xy] = T_0[y]x^\dagger = x_R^\dagger T_0[y]. \tag{7.14}$$

So in general,

$$\boxed{\begin{aligned} T_0 x_L &= x_R^\dagger T_0, \\ T_0 x_L^\dagger &= x_R T_0, \\ T_0 x_R &= x_L^\dagger T_0, \\ T_0 x_R^\dagger &= x_L T_0 \end{aligned}} \tag{7.15}$$

(the last three of these equations are derived in a like manner). Furthermore,

$$T_0^2 = (2P_0)^2 - 4P_0 + 1 = 4P_0 - 4P_0 + 1 = 1. \tag{7.16}$$

Therefore,

$$\boxed{\begin{aligned} x_L &= T_0 x_R^\dagger T_0, \\ x_L^\dagger &= T_0 x_R T_0, \\ x_R &= T_0 x_L^\dagger T_0, \\ x_R^\dagger &= T_0 x_L T_0. \end{aligned}} \tag{7.17}$$

Using $T_0^2 = 1$ we derive

$$\begin{aligned} T_0 e_{Lab...c} T_0 &= T_0 e_{La} e_{Lb} ... e_{Lc} T_0 \\ &= T_0 e_{La} T_0 T_0 e_{Lb} T_0 ... T_0 e_{Lc} T_0 \\ &= e_{Ra}^\dagger e_{Rb}^\dagger ... e_{Rc}^\dagger \\ &= (e_{Rc} ... e_{Rb} e_{Ra})^\dagger \\ &= (e_{Rab...c})^\dagger. \end{aligned} \tag{7.18}$$

Therefore, for $a, b, c \in \{1, ..., 7\}$ distinct,

$$\boxed{\begin{aligned} T_0 e_{La} T_0 &= -e_{Ra}, \\ T_0 e_{Lab} T_0 &= -e_{Rab}, \\ T_0 e_{Labc} T_0 &= +e_{Rabc}. \end{aligned}} \tag{7.19}$$

It should be clear that the map

$$A \to T_0 A T_0$$

on $A \in \mathbf{O}_L$ is an $\mathbf{O}_L$ automorphism.

In the last equation above, consider the special cases

$$I_a = e_{L(3+a)(6+a)(5+a)} \tag{7.20}$$

(see 2.46), indices from 1 to 7, modulo 7. (Note, for example, that

$$P_0 = \frac{1}{8}(1 + \sum_1^7 I_a).)$$

The I_a mutually commute:

$$I_a I_b = I_b I_a \tag{7.21}$$

(in fact, $I_a P_0 = P_0 I_a = P_0$). Combining 7.18,19,21 yields

$$T_0 I_a T_0 = I_a T_0^2 = I_a.$$

Therefore, if $e_{Labc}[1] = \pm 1$ (so $e_{Labc} = \pm I_d$, for some $d \in \{1, ..., 7\}$), then

$$T_0 e_{Labc} T_0 = e_{Rabc} = e_{Labc}. \tag{7.22}$$

Equation 7.22 makes it clear that in the definitions of the P_a in 7.1, it was unnecessary to specify left-action. Specifically,

$$\begin{aligned} P_0 &= \tfrac{1}{8}(1 + e_{L476} + e_{L517} + e_{L621} + e_{L732} + e_{L143} + e_{L254} + e_{L365}) \\ &= \tfrac{1}{8}(1 + e_{R476} + e_{R517} + e_{R621} + e_{R732} + e_{R143} + e_{R254} + e_{R365}). \end{aligned} \tag{7.23}$$

TWO IDENTITIES

One of the principal identities of this section is the following:

$$\boxed{\begin{aligned}(xy)_L &= x_R T_0 y_R - y_R^\dagger T_0 x_R^\dagger + x_R y_R \\ &= 2x_R P_0 y_R - y_R^\dagger T_0 x_R^\dagger \\ &= 2x_R P_0 y_R - 2y_R^\dagger P_0 x_R^\dagger + y_R^\dagger x_R^\dagger;\end{aligned}} \tag{7.24}$$

from which it immediately follows that

$$\boxed{\begin{aligned}(xy)_R &= y_L T_0 x_L - x_L^\dagger T_0 y_L^\dagger + y_L x_L \\ &= 2y_L P_0 x_L - x_L^\dagger T_0 y_L^\dagger \\ &= 2y_L P_0 x_L - 2x_L^\dagger P_0 y_L^\dagger + x_L^\dagger y_L^\dagger.\end{aligned}} \tag{7.25}$$

Therefore, for $a, b \in \{1, ..., 7\}$ distinct, it is now not difficult to prove

$$\boxed{[e_a e_b]_L = -e_{Rb} e_{Lab} e_{Rb}.} \tag{7.26}$$

Proof of 7.24.

We may prove these identities by considering their pieces. Since $e_1 = e_2 e_6$, it is clear that $(e_2 e_6)_L = e_{L1}$. Consider the element

$$P_0 + P_1 + P_2 + P_6 = \frac{1}{2}(1 + e_{L621}). \tag{7.27}$$

Being the sum of orthogonal idempotents, this too is an idempotent. 7.27 implies

$$e_{L621} = 2(P_0 + P_1 + P_2 + P_6) - 1 = -T_0 T_1 T_2 T_6.$$

Therefore,

$$
\begin{aligned}
-(e_2e_6)_L &= -e_{L1} \\
&= -e_{L26}e_{L621} \\
&= e_{L26}T_0T_1T_2T_6 \\
&= T_0e_{L2}T_0T_1e_{L6}T_0 \\
&= T_0e_{L2}[1-2P_0-2P_1]e_{L6}T_0 \\
&= T_0e_{L2}[(\tfrac{1}{2}-2P_0)+(\tfrac{1}{2}-2P_1)]e_{L6}T_0 \\
&= T_0e_{L2}(\tfrac{1}{2}-2P_0)e_{L6}T_0+T_0e_{L2}(\tfrac{1}{2}-2P_1)e_{L6}T_0 \\
&= T_0e_{L2}(\tfrac{1}{2}-2P_0)e_{L6}T_0-T_0e_{L6}(\tfrac{1}{2}-2P_0)e_{L2}T_0 \\
&= T_0e^{\dagger}_{L2}(2P_0-\tfrac{1}{2})e_{L6}T_0-T_0e^{\dagger}_{L6}(2P_0-\tfrac{1}{2})e_{L2}T_0 \\
&= e_{R2}(2P_0-\tfrac{1}{2})e^{\dagger}_{R6}-e_{R6}(2P_0-\tfrac{1}{2})e^{\dagger}_{R2} \\
&= e_{R2}T_0e^{\dagger}_{R6}-e_{R6}T_0e^{\dagger}_{R2}+\tfrac{1}{2}(e_{R2}e^{\dagger}_{R6}-e_{R6}e^{\dagger}_{R2}) \\
&= -[e_{R2}T_0e_{R6}-e^{\dagger}_{R6}T_0e^{\dagger}_{R2}+e_{R2}e_{R6}].
\end{aligned}
\tag{7.28}
$$

This proves 7.24 for the case $x = e_a$, $y = e_b$, $a, b \in \{1, ..., 7\}$, $a \neq b$.

Likewise, for $a \in \{1, ..., 7\}$,

$$\begin{aligned} (1e_a)_L &= e_{La} \\ &= T_0 e^\dagger_{Ra} T_0 \\ &= T_0 T_a e^\dagger_{Ra} \\ &= -(2P_0 + 2P_a - 1)e^\dagger_{Ra} \\ &= -(2P_a - \tfrac{1}{2})e^\dagger_{Ra} - (2P_0 - \tfrac{1}{2})e^\dagger_{Ra} \\ &= -e^\dagger_{Ra}(2P_0 - \tfrac{1}{2}) - (2P_0 - \tfrac{1}{2})e^\dagger_{Ra} \\ &= 1_R(2P_0 - \tfrac{1}{2})e_{Ra} - e^\dagger_{Ra}(2P_0 - \tfrac{1}{2})1^\dagger_R \\ &= 1_R T_0 e_{Ra} - e^\dagger_{Ra} T_0 1^\dagger_R + 1_R e_{Ra}. \end{aligned} \tag{7.29}$$

Clearly, therefore, we may reverse 1 and e_a to arrive at

$$(e_a 1)_L = e_{La} = e_{Ra} T_0 1_R - 1^\dagger_R T_0 e^\dagger_{Ra} + e_{R1} 1_R. \tag{7.30}$$

This proves 7.24 in the case $x = 1$ and $y = e_a$, and visa versa.

Penultimately, since $e_{Ra}e_{Ra} = e^\dagger_{Ra}e^\dagger_{Ra} = -1_R = -1_L$,

$$\begin{aligned} (e_a e_a)_L = -1_L &= (e_{Ra}e_{Ra} - e^\dagger_{Ra}e^\dagger_{Ra}) + e_{Ra}e_{Ra} \\ &= (e_{Ra}T_0 e_{Ra} - e^\dagger_{Ra}T_0 e^\dagger_{Ra}) + e_{Ra}e_{Ra}, \end{aligned} \tag{7.31}$$

since $e_{Ra}T_0e_{Ra} - e^\dagger_{Ra}T_0e^\dagger_{Ra} = 0$.

Finally it should be clear that

$$((1)(1))_L = 1_R T_0 1_R - 1^\dagger_R T_0 1^\dagger_R + 1_R 1_R, \tag{7.32}$$

which completes the proof of 7.24, and by extension 7.25.

THE MOUFANG IDENTITIES

These results can be used to prove the Moufang identities,

$$(xy)(ax) = x(ya)x, \tag{7.33}$$

and

$$(xax)y = x(a(xy)), \tag{7.34}$$

which can be written in adjoint notation as

$$\boxed{(xy)_L x_R = x_L x_R y_L = x_R x_L y_L,} \tag{7.35}$$

and

$$\boxed{y_R x_L x_R = y_R x_R x_L = x_L (xy)_R.} \tag{7.36}$$

First observe that as $Re(ax) = Re(xa)$ for all $a, x \in \mathbf{O}$,

$$P_0 x_L = P_0 x_R \tag{7.37}$$

in general. Also, $x Re(a) = Re(a) x$, so

$$x_L P_0 = x_R P_0 \tag{7.38}$$

in general.

The associator is defined by

$$\boxed{(y, a, x) = y(ax) - (ya)x = (y_L x_R - x_R y_L)[a].} \tag{7.39}$$

It is well-known that the associator is antisymmetric in y, a and x, and that it has no real part. Therefore,

$$P_0(y_L x_R - x_R y_L)[a] = Re(y, a, x) = 0,$$

which implies

$$P_0(y_L x_R - x_R y_L) = 0. \tag{7.40}$$

With these results in hand we are ready to prove the Moufang identities.

Together with 7.24, and starting with the lefthand side of 7.35,

$$\begin{aligned}(xy)_L x_R &= x_R T_0 y_R x_R - y_R^\dagger T_0 x_R^\dagger x_R + x_R y_R x_R \\ &= 2x_R P_0 y_R x_R - y_R^\dagger T_0 x_R^\dagger x_R \\ &= 2x_R P_0 y_R x_R - y_R^\dagger T_0 \|x\|^2 \\ &= 2x_R P_0 y_R x_R - T_0 y_L \|x\|^2 \\ &= 2x_R P_0 y_R x_R - T_0 \|x\|^2 y_L \\ &= 2x_R P_0 y_R x_R - T_0 x_R^\dagger x_R y_L \qquad (7.41) \\ &= 2x_R P_0 y_R x_R - x_L T_0 x_R y_L \\ &= 2x_R P_0 y_R x_R - 2x_L P_0 x_R y_L + x_L x_R y_L \\ &= 2x_L P_0 y_L x_R - 2x_L P_0 x_R y_L + x_L x_R y_L \\ &= 2x_L P_0 (y_L x_R - x_R y_L) + x_L x_R y_L \\ &= x_L x_R y_L,\end{aligned}$$

which is the Moufang identity 7.35. Equation 7.36 can be proven similarly.

To finish this section we may as well prove that the real part of the associator is zero. Note that

$$\begin{aligned}Re(x,y,z) &= Re(x(yz) - (xy)z) \\ &= Re(x_L y_L[z] - (xy)_L[z]) \\ &= P_0(x_L y_L[z] - (xy)_L[z]).\end{aligned}$$

Therefore proving $Re(x,y,z) = 0$ is equivalent to proving

$$P_0 x_L y_L - P_0 (xy)_L = 0. \qquad (7.42)$$

Using all the tools developed in this section, together with the somewhat obvious

$$P_0 y_L P_0 = y^0 P_0 = \frac{1}{2}(y_L + y_L^\dagger) P_0, \qquad (7.43)$$

equation 7.42 is not difficult to prove:

$$\begin{aligned}
P_0 x_L y_L - P_0(xy)_L &= P_0 x_L y_L - P_0(xy)_R \\
&= P_0 x_L y_L - 2P_0 y_L P_0 x_L + P_0 x_L^\dagger T_0 y_L^\dagger \\
&= P_0 x_L y_L - 2y^0 P_0 x_L + P_0 T_0 x_R y_L^\dagger \\
&= P_0 x_L y_L - 2y^0 P_0 x_L + P_0 x_R y_L^\dagger \\
&= P_0 x_L y_L - 2y^0 P_0 x_L + P_0 x_L y_L^\dagger \\
&= P_0 x_L (y_L + y_L^\dagger) - 2y^0 P_0 x_L \\
&= P_0 x_L (2y^0) - 2y^0 P_0 x_L \\
&= 0.
\end{aligned}$$

7.2. Spheres and Lie Algebras.

The parallelizable spheres are S^0, S^1, S^3, and S^7. Let $\mathbf{K}$ be a division algebra, and let $\mathbf{K}_u$ be the set of elements of $\mathbf{K}$ of unit length. Then

$$\boxed{\mathbf{R}_u \simeq S^0,\ \mathbf{C}_u \simeq S^1,\ \mathbf{Q}_u \simeq S^3,\ \mathbf{O}_u \simeq S^7.} \tag{7.44}$$

The trivial sphere $\mathbf{R}_u \simeq S^0$ consists of the two points $\{-1, 1\}$ and is of little interest. The others are easily generated by exponentiation. Let θ, $\phi^j, j = 1, 2, 3$, and $\vartheta^a, a = 1, ..., 7$, be real, then

$$\boxed{S^1 = \{e^{i\theta}\},\ S^3 = \{e^{\sum \phi^j q_j}\},\ S^7 = \{e^{\sum \vartheta^a e_a}\}.} \tag{7.45}$$

Parallelizing tangent vector fields are easily constructed on these spheres given the representations 7.45. (I shall leave it to the readers, should they so choose, to cloak these simple structures in the elaborate symbollism of differential geometry.) At each point of $S^h, h = 1, 3, 7$, the h basis vectors

$$\{ie^{i\theta}\},\ \{q_i e^{\sum \phi^j q_j}\},\ \{e_b e^{\sum \vartheta^a e_a}\} \tag{7.46}$$

span the tangent space at the point, and they clearly map smoothly around the spheres as the values of the parameters change. To prove these vector directions in **C**-, **Q**-, and **O**-space are tangent to the spheres, we need merely prove they are perpendicular to the points to which they are attached. That is, let u be an element of $S^h, h = 1,3,7$, and let c_k denote i, q_i, or e_a, respectively. Then

$$\begin{aligned} < u, c_k u > &= \tfrac{1}{2}(u^\dagger(c_k u) + (c_k u)^\dagger u) \\ &= \tfrac{1}{2}(u^\dagger c_k u + u^\dagger c_k^\dagger u) \\ &= \tfrac{1}{2}(u^\dagger c_k u - u^\dagger c_k u) = 0, \end{aligned} \tag{7.47}$$

thereby proving that the vectors 7.46 are tangent to the spheres 7.45 (note that $u^\dagger(c_k u) = (u^\dagger c_k)u$ even in the nonassociative **O** case).

S^3

Let

$$X = e^{\sum \phi^j q_j} = X^0 + \sum X^j q_j \in S^3. \tag{7.48}$$

Then $\mathcal{T}_X$, the tangent space at X, is spanned by the orthonormal basis

$$\{q_i X : i = 1,2,3\}. \tag{7.49}$$

Any infinitesimal tangential movement from X can be represented by an action of the form

$$X \to X + \sum \delta^i q_i X. \tag{7.50}$$

Consider a pair of such transformations:

$$\begin{aligned} X \to X + \delta^2 q_2 X &\to X + \delta^2 q_2 X + \delta^1 q_1 (X + \delta^2 q_2 X) \\ &= X + \delta^2 q_2 X + \delta^1 q_1 X + \delta^1 \delta^2 q_3 X, \end{aligned} \tag{7.51}$$

where $q_3 X \in \mathcal{T}_X$. Reversing the order of these transformations leads to

$$X \to X + \delta^2 q_2 X + \delta^1 q_1 X - \delta^1 \delta^2 q_3 X. \tag{7.52}$$

Therefore, as expected, these two transformations do not commute.

Any infinitesimal tangential movement from $X \in S^3$ should be representable in the form 7.50. In particular, the elements

$$\{q_{Li}, q_{Rj}\} \tag{7.53}$$

form a basis for the Lie algebra $so(4)$, the infinitesimal rotations on the 3-sphere. Rotations along the first three directions are easily representable in the form 7.50. That is,

$$X \to X + \sum \delta^i q_{Li}[X] = X + \sum \delta^i q_i X. \tag{7.54}$$

Along the remaining three directions it is only slightly more difficult. Using the fact that $X^\dagger X = 1$, we derive:

$$\begin{aligned} X &\to X + \sum \delta^i q_{Ri}[X] = X + \sum \delta^i X q_i \\ &= X + \sum (\delta^i X q_i X^\dagger) X = X + \sum \bar{\delta}^i(X) q_i X, \end{aligned} \tag{7.55}$$

where the infinitesimal $\bar{\delta}^i(X)$ are position dependent, and we have made use of the fact that $q_i \to X q_i X^\dagger$ is an $SO(3)$ rotation of q_i, linear in $\{q_1, q_2, q_3\}$.

In the S^3 case the Lie algebra of the isometry group of the manifold, $so(4)$, is generated by the elements $\{q_{Li}, q_{Rj}\}$. The subset $\{q_{Li}\}$ closes under commutation, as does the subset $\{q_{Rj}\}$, and each is isomorphic to $so(3) \simeq su(2) = spin(3)$. Actions arising from the former subalgebra, $q_{Li}[X] = q_i X$, generate bases for the chosen $\mathcal{T}_X$. (Of course, one obvious other choice for a basis for $\mathcal{T}_X$ would be elements of the form $X q_i$, which also parallelize the sphere.)

Note that because the $\{q_{Li}\}$ close under commutation, the Lie group of all elements of the form

$$\{e^{\sum \phi^j q_{Lj}}\} \simeq S^3 \simeq SU(2). \tag{7.56}$$

Finally, the set of elements

$$\{q_{Li} - q_{Ri}\}, \tag{7.57}$$

with action on $X \in S^3$,

$$X \to q_{Li}[X] - q_{Ri}[X] = q_i X - X q_i, \tag{7.58}$$

leaves the identity direction invariant. Exponentiated they generate the group $SO(3)$ on the remaining purely hypercomplex coordinates, which is the automorphism group of $\mathbf{Q}$. The subgroup of this $SO(3)$ that leaves an additional $\mathbf{Q}$ direction invariant, in particular q_3, is $SO(2)$, generated in that case by the single element

$$\{q_{L3} - q_{R3}\}. \tag{7.59}$$

The automorphism on **Q** defined by

$$q_1 \to q_2,\ q_2 \to q_1,\ q_3 \to -q_3,$$

is an element of this $SO(2)$. This is index doubling with a twist (index doubling is an antiautomorphism on **Q** that takes $q_3 \to q_3$).

S^7

Now let

$$X = e^{\sum \vartheta^a e_a} = X^0 + \sum_1^7 X^a e_a \in S^7. \tag{7.60}$$

$\mathcal{T}_X$, the tangent space at X, is spanned by the orthonormal basis

$$\{e_a X : a = 1, ..., 7\}, \tag{7.61}$$

and again any infinitesimal tangential movement from X can be represented by an action of the form:

$$X \to X + \sum \delta^a e_a X = X + \sum \delta^a e_{La}[X]. \tag{7.62}$$

In the quaternion case consecutive transformations of the form 7.50 were also of that form. Here things are different. Consider the consecutive transformations below:

$$\begin{aligned} X \to X + \delta^2 e_2 X \quad &\to \quad X + \delta^2 e_2 X + \delta^1 e_1(X + \delta^2 e_2 X) \\ &= \quad X + \delta^2 e_2 X + \delta^1 e_1 X + \delta^1\delta^2 e_1(e_2 X) \qquad (7.63) \\ &= \quad X + (\delta^2 e_{L2}[X] + \delta^1 e_{L1}[X]) + \delta^1\delta^2 e_{L12}[X]. \end{aligned}$$

In reverse order

$$X \to X + (\delta^2 e_{L2}[X] + \delta^1 e_{L1}[X]) - \delta^1\delta^2 e_{L12}[X]. \tag{7.64}$$

The problem here is that whereas the $\{q_{Li}\}$ close under commutation, hence give rise to a group manifold, the $\{e_{La}\}$ do not. Therefore, even though the elements

$$X_L = e^{\sum \vartheta^a e_{La}} = X^0 + \sum_1^7 X^a e_{La} \tag{7.65}$$

constitute another copy of S^7, the set of such elements is not a group manifold (S^7 is the unique parallelizable manifold not also a group manifold).

The Lie algebra of the isometry group of S^7 is $so(8)$, a basis for which in this context consists of the 28 elements

$$\{e_{La}, e_{Lab}\}. \tag{7.66}$$

The action of the e_{La} on $X \in S^7$ generates the given basis for $\mathcal{T}_X$:

$$X \rightarrow e_{La}[X] = e_a X. \tag{7.67}$$

As we have seen, however, and in distinction to the quaternion case, these actions do not close under commutation.

$$\frac{1}{2}[e_{La}, e_{Lb}] = e_{Lab}, \ a \neq b. \tag{7.68}$$

That is, they generate the remainder of the $so(8)$ Lie algebra.

The set of elements of the form $\{e_{Lab}\}$ does close under commutation, constituting a representation of the Lie algebra $so(7)$. It is well-known that

$$SO(8)/SO(7) \simeq S^7. \tag{7.69}$$

Highlighting this holomorphism is the fact that the set of X_L in 7.65 is also S^7. In that case, however, it is not so easy algebraically to parallelize the sphere, as

$$e_{La}(X^0 + \sum X^b e_{Lb}) = X^0 e_{La} + \sum X^b e_{Lab}, \tag{7.70}$$

where $e_{Laa} = -1_L$. That is, the set of $\{e_{La}\}$, unlike the $\{e_a\}$, do not close under multiplication, so $e_{La}X_L$ does not reside in the same space as X_L.

Back to $\{X\} = S^7$. The infinitesimal $so(7)$ transformations arising from $\{e_{Lab}\}$ must be representable in terms of movements along the directions $e_a X$, since these span $\mathcal{T}_X$. In fact, using the fact that $(YX^\dagger)X = Y$ for all $Y \in \mathbf{O}$, and $X \in S^7$,

$$e_{Lab}[X] = e_a(e_b X) = ((e_a(e_b X))X^\dagger)X \tag{7.71}$$

(see [8]), where we set

$$((e_a(e_b X))X^\dagger) = \sum_{c=1}^{7} \bar{\delta}^c_{ab}(X) e_c. \tag{7.72}$$

That is, $(e_a(e_bX))X^\dagger$ has no real part. This is easily demonstrated. Since $e_a(e_bX) = e_{Lab}[X]$ is an infinitesimal $so(8)$ rotation of X, the term $e_{Lab}[X] \in \mathcal{T}_X$ must be perpendicular to X. Since in general, for all $Y, Z \in \mathbf{O}$,

$$\begin{aligned} P_0[YZ] &= P_0[(YZ)^\dagger] \\ &= \tfrac{1}{2}P_0[YZ + (YZ)^\dagger] \\ &= \tfrac{1}{2}P_0[Y(Z^\dagger)^\dagger + Z^\dagger Y^\dagger] \\ &= \tfrac{1}{2}P_0[2 < Y, Z^\dagger >] \\ &= < Y, Z^\dagger >, \end{aligned}$$

then

$$\begin{aligned} Re((e_a(e_bX))X^\dagger) &= P_0[(e_{Lab}[X])X^\dagger] \\ &= < e_{Lab}[X], X > \\ &= 0. \end{aligned} \tag{7.73}$$

This justifies writing $(e_a(e_bX))X^\dagger$ in the form 7.72.

[Note: 7.72 can also be written

$$\boxed{(e_a(e_bX))X^\dagger = (e_aX)(X^\dagger e_b).} \tag{7.74}$$

This is equivalent to

$$X_R^\dagger e_{La} X_R = (e_aX)_L X_L^\dagger.$$

Using 7.24 we note that

$$\begin{aligned} (e_aX)_L X_L^\dagger &= 2e_{Ra}P_0X_RX_L^\dagger - X_R^\dagger T_0 e_{Ra}^\dagger X_L^\dagger \\ &= 2e_{Ra}P_0X_LX_L^\dagger - X_R^\dagger e_{La}X_RT_0 \\ &= 2e_{Ra}P_0 - 2X_R^\dagger e_{La}X_RP_0 + X_R^\dagger e_{La}X_R \\ &= 2e_{Ra}P_0 - 2e_{La}P_0 + X_R^\dagger e_{La}X_R \\ &= X_R^\dagger e_{La}X_R, \end{aligned}$$

this proving 7.74 (a couple of steps are left for the reader to justify). Note that for any $X \in S^7$, the product $(e_a X)(X^\dagger e_b)$ defines a new octonion multiplication, which in [8] is called the X-product . That this gives rise to an altered octonion multiplication is due to the nonassociativity of $\mathbf{O}$, for in any other division algebra the X and $X^\dagger$ would immediately cancel. This new product was used in [8] to give S^7 a group-like structure.]

To illustrate another of the tools applicable in this context, we'll calculate the coefficients $\bar{\delta}^c_{ab}(X)$ for the case $a = 1$, $b = 2$. That is,

$$((e_1(e_2X))X^\dagger) = \sum_{c=1}^{7} \bar{\delta}^c_{12}(X)e_c. \tag{7.75}$$

Consider the table below of actions of 1_L and e_{La} on X:

	1	e_1	e_2	e_3	e_4	e_5	e_6	e_7
1_L	X^0	X^1	X^2	X^3	X^4	X^5	X^6	X^7
e_{L1}	$-X^1$	X^0	$-X^6$	$-X^4$	X^3	$-X^7$	X^2	X^5
e_{L2}	$-X^2$	X^6	X^0	$-X^7$	$-X^5$	X^4	$-X^1$	X^3
e_{L3}	$-X^3$	X^4	X^7	X^0	$-X^1$	$-X^6$	X^5	$-X^2$
e_{L4}	$-X^4$	$-X^3$	X^5	X^1	X^0	$-X^2$	$-X^7$	X^6
e_{L5}	$-X^5$	X^7	$-X^4$	X^6	X^2	X^0	$-X^3$	$-X^1$
e_{L6}	$-X^6$	$-X^2$	X^1	$-X^5$	X^7	X^3	X^0	$-X^4$
e_{L7}	$-X^7$	$-X^5$	$-X^3$	X^2	$-X^6$	X^1	X^4	X^0

(7.76)

That is, for example,

$$e_{L1}[X] = e_1X = -X^1e_0 + X^0e_1 - X^6e_2 - X^4e_3 + X^3e_4 - X^7e_5 + X^2e_6 + X^5e_7.$$

So the zeroth row gives the components of $X \in S^7$, and the remaining rows the components of $e_aX \in \mathcal{T}_X$.

Let M_X denote the 8×8 matrix of coefficients in 7.76. Since the rows constitute an orthonormal basis for $\mathbf{R}^8$, it is clear that

$$M_X^\dagger = M_X^{-1}. \tag{7.77}$$

Let $\vec{X}_a, a = 0, ..., 7$, be the ath row of M_X, an $\mathbf{R}^8$ unit vector.

Observe that

$$e_{L12}[X] = -X^6e_0 - X^2e_1 + X^1e_2 + X^5e_3 - X^7e_4 - X^3e_5 + X^0e_6 + X^4e_7.$$

Define

$$\vec{X}_{12} = [-X^6, -X^2, X^1, X^5, -X^7, -X^3, X^0, X^4] \in \mathbf{R}^8.$$

7.75 implies

$$((e_{L12}[X])X^\dagger)X = e_{L12}[X] = \sum_{c=1}^{7} \bar{\delta}^c_{12}(X)e_c X. \tag{7.78}$$

Define

$$\vec{\delta}_{12}(X) = [\bar{\delta}^0_{12}, ..., \bar{\delta}^7_{12}]. \tag{7.79}$$

Then 7.78 is equivalent to

$$\vec{X}_{12} = \sum_{b=0}^{7} \bar{\delta}^b_{12}(X)\vec{X}_b = \vec{\delta}_{12}(X)M_X.$$

Therefore, $\vec{\delta}_{12}(X) = \vec{X}_{12}M_X^{-1}$, or

$$(\vec{\delta}_{12}(X))^\dagger = M_X \vec{X}^\dagger_{12}, \tag{7.80}$$

where $(\vec{\delta}_{12}(X))^\dagger$ and $\vec{X}^\dagger_{12}$ are column vectors. This implies

$$\begin{aligned}
&((e_{L12}[X])X^\dagger) = (e_1 X)(X^\dagger e_2)\\
&= ((X^0)^2 + (X^1)^2 + (X^2)^2 + (X^6)^2 - (X^3)^2 - (X^4)^2 - (X^5)^2 - (X^7)^2)e_6\\
&+2(X^0X^5 + X^1X^7 - X^2X^4 + X^3X^6)e_3\\
&+2(-X^0X^7 + X^1X^5 + X^2X^3 + X^4X^6)e_4\\
&+2(-X^0X^3 - X^1X^4 - X^2X^7 + X^5X^6)e_5\\
&+2(X^0X^4 - X^1X^3 + X^2X^5 + X^7X^6)e_7.
\end{aligned} \tag{7.81}$$

That is, $\bar{\delta}^b_{12}(X) = 0$ for $b = 0, 1, 2$, and is given by the terms in parentheses for $b = 3, 4, 5, 6, 7$.

All the remaining $e_{Lab}[X]$ can be computed by index doubling and cycling on the index set $\{1, ..., 7\}$ as they appear in 7.81. For example, to calculate

$e_{L74}[X]$ we note that $7 = 4(1)+3 \bmod 7$, and $4 = 4(2)+3 \bmod 7$. So perform the operation $a \to 4(a)+3$ (modulo 7) on all $a \in \{1, ..., 7\}$ appearing in 7.81. This has the effect of taking

$$1 \to 7 \to 3 \to 1,$$
$$2 \to 4 \to 5 \to 2,$$
$$6 \to 6.$$

We conclude that

$$\begin{aligned}
&((e_{L74}[X])X^\dagger) \\
&= ((X^0)^2 + (X^7)^2 + (X^4)^2 + (X^6)^2 - (X^1)^2 - (X^5)^2 - (X^2)^2 - (X^3)^2)e_6 \\
&+2(X^0X^2 + X^7X^3 - X^4X^5 + X^6X^1)e_1 \\
&+2(-X^0X^3 + X^7X^2 + X^4X^1 + X^6X^5)e_5 \\
&+2(-X^0X^1 - X^7X^5 - X^4X^3 + X^6X^2)e_2 \\
&+2(+X^0X^5 - X^7X^1 + X^4X^2 + X^6X^3)e_3.
\end{aligned} \tag{7.82}$$

As pointed out in **[8]**, the $\bar{\delta}^c_{ab}(X)$ are the components of the parallelizing torsion of $S^7 = \{X\}$, which, unlike the case for group manifolds, is position dependent (this is a direct consequence of the nonassociativity of $\mathbf{O}$).

As we did for the quaternion case, we can construct the subalgebra of $so(8)$ that leaves the identity invariant. A basis for this consists of the 21 distinct elements of the form

$$so(7) \to \{e_{La} - e_{Lbc}, e_{Lbc} - e_{Lrs} : e_a = e_b e_c = e_r e_s\}. \tag{7.83}$$

This set generates a vector representation of $SO(7)$ via exponentiation, rotating the seven imaginary dimensions of $\mathbf{O}$.

The subalgebra of $so(7)$ leaving another octonion direction invariant, in particular e_7, is $so(6)$, with basis

$$\begin{aligned}
so(6) \to \{&e_{La} - e_{Lbc}, e_{L7} - e_{L23}, e_{Lpq} - e_{Lrs} : \\
&b \text{ or } c = 7, e_a = e_b e_c,\ e_p e_q = e_r e_s, a, p, q, r, s \neq 7\}.
\end{aligned} \tag{7.84}$$

The automorphism group of **O** is G_2, with a Lie algebra basis a subset of that of $so(7)$ in 7.83:

$$LG_2 \to \{e_{Lab} - e_{Lcd} : e_a e_b = e_c e_d\}. \tag{7.85}$$

The intersection of LG_2 with the $so(6)$ in 7.84 is $su(3)$:

$$su(3) \to \{e_{Lpq} - e_{Lrs} : e_p e_q = e_r e_s, p, q, r, s \neq 7\}. \tag{7.86}$$

Note that $so(7)$ has another basis:

$$so(7) \to \{e_{Lab}\}. \tag{7.87}$$

However, these do not leave any elements of **O** invariant, and via exponentiation generate a half spinor representation of $Spin(7)$ acting on $\mathbf{R}^8 \simeq \mathbf{O}$.

The subalgebra

$$so(6) \to \{e_{Lpq} : p, q \neq 7\} \tag{7.88}$$

also leaves no elements of **O** invariant, and via exponentiation generates a representation of $SU(4)$ acting on $\mathbf{C}^4 \simeq \mathbf{O}$. $Spin(7)/G_2$ is also holomorphic to S^7.

The intersection of the vector $so(6)$ of 7.84, and spinor $so(6)$ of 7.88, is again $su(3)$.

SPHERE FIBRATIONS

Let u_1, u_2 be elements of the sphere $S^h, h = 1, 3$, or 7, so u_1 and u_2 are unit elements of **C**, **Q**, or **O**. Let φ be real, then

$$S^{2h+1} = \{\begin{bmatrix} u_1 cos\varphi \\ u_2 sin\varphi \end{bmatrix}\}. \tag{7.89}$$

The Hermitian matrix

$$\begin{aligned}
&\begin{bmatrix} u_1 cos\varphi \\ u_2 sin\varphi \end{bmatrix} [\, u_1^\dagger cos\varphi \quad u_2^\dagger sin\varphi \,] \\
&= \begin{bmatrix} cos^2\varphi & u_1 u_2^\dagger cos\varphi sin\varphi \\ u_2 u_1^\dagger cos\varphi sin\varphi & sin^2\varphi \end{bmatrix} \\
&= \begin{bmatrix} \frac{1}{2}(1 + cos2\varphi) & \frac{1}{2} u_1 u_2^\dagger sin2\varphi \\ \frac{1}{2} u_2 u_1^\dagger sin2\varphi & \frac{1}{2}(1 - cos2\varphi) \end{bmatrix} \\
&= \tfrac{1}{2}\epsilon + \tfrac{1}{2}[\alpha cos2\varphi + \beta(< u_1, u_2 > sin2\varphi) + \omega(u_1 \times u_2 sin2\varphi)]
\end{aligned} \tag{7.90}$$

defines an element of S^{h+1} (the term inside the brackets in the last line of 7.90 lies on the unit sphere S^{h+1}), hence also defines a map from $S^{2h+1} \to S^{h+1}$. Note,

$$< u_1, u_2 >= \frac{1}{2}(u_1 u_2^\dagger + u_2 u_1^\dagger),$$

$$u_1 \times u_2 = \frac{1}{2}(u_1 u_2^\dagger - u_2 u_1^\dagger).$$

Let $u \in S^h$, and define the map

$$u \to \begin{bmatrix} u \\ 0 \end{bmatrix} \tag{7.91}$$

of $S^h \to S^{2h+1}$, which, via the map 7.90,

$$\begin{bmatrix} u \\ 0 \end{bmatrix} \to \begin{bmatrix} u \\ 0 \end{bmatrix} [u^\dagger \; 0] = \begin{bmatrix} 1 & 0 \\ 0 & 0 \end{bmatrix}, \tag{7.92}$$

a single point in S^{h+1}. In this way, via rotations of S^{2h+1} the map 7.87 can be generalized and lead via 7.88 to any point in S^{h+1}. This is summarized by the exact sequence

$$1 \to S^h \to S^{2h+1} \to S^{h+1} \to 1 \tag{7.93}$$

which defines an S^h fibre over each point of S^{h+1}.

The parallelizable spheres and these sphere fibrations are the only ones of their kind, a part of the select web of mathematical objects surrounding the division algebras, and,I believe, an inevitable part of the design of reality.

7.3. Triality.

At the end of section 2.5 triality was briefly mentioned. We'll develop the idea more carefully here as another application of the ideas underlying the physical application of chapters 3 to 6.

We begin with the Clifford algebra $\mathbf{R}_{8,0}$, with 1-vector basis

$$\beta,\ e_{La}\omega. \tag{7.94}$$

This acts on the spinor space $\mathbf{O}^2$ of octonion doublets. Let $x, y, z \in \mathbf{O}$, with $x_L, y_L, z_L, x_R, y_R, z_R$ defined as in sections 7.1 and 7.2. Define

$$X_L = x^0\beta + \sum_1^7 x^a e_{La}\omega = \begin{bmatrix} 0 & x_L \\ x_L^\dagger & 0 \end{bmatrix}. \tag{7.95}$$

On $\mathbf{O}$ define the trilinear form

$$\boxed{\begin{aligned} <x, y, z> &= \tfrac{1}{2} < \begin{bmatrix} y^\dagger \\ z \end{bmatrix}, X_L \begin{bmatrix} y^\dagger \\ z \end{bmatrix} > \\ &= \tfrac{1}{2} < \begin{bmatrix} y^\dagger \\ z \end{bmatrix}, \begin{bmatrix} x_L[z] \\ x_L^\dagger[y^\dagger] \end{bmatrix} > \\ &= \tfrac{1}{2}(< y^\dagger, x_L[z] > + < z, x_L^\dagger[y^\dagger] > \\ &= < y^\dagger, xz > \\ &= P_0[y(xz)] = P_0[(yx)z], \end{aligned}} \tag{7.96}$$

since

$$< z, x_L^\dagger[y^\dagger] >=< x_L[z], y^\dagger >=< y^\dagger, x_L[z] > .$$

The 2-vector basis of $\mathbf{R}_{8,0}$ corresponding to 7.94 consists of the elements

$$-e_{La}\alpha, \quad e_{Lab}\epsilon. \tag{7.97}$$

Under commutation these elements generate the Lie algebra $so(8)$, and under exponentiation $Spin(8)$. Let U_L be such a group element. The action

$$X_L \rightarrow U_L X_L U_L^{-1} \tag{7.98}$$

is $SO(8)$, and

$$\begin{bmatrix} y^\dagger \\ z \end{bmatrix} \rightarrow U_L \begin{bmatrix} y^\dagger \\ z \end{bmatrix} \tag{7.99}$$

is $Spin(8)$. It is clear that the trilinear form 7.96 is invariant under these actions:

$$
\begin{aligned}
<x,y,z> &\rightarrow \tfrac{1}{2} < U_L \begin{bmatrix} y^\dagger \\ z \end{bmatrix}, U_L X_L U_L^{-1} U_L \begin{bmatrix} y^\dagger \\ z \end{bmatrix} > \\
&= \tfrac{1}{2} < U_L \begin{bmatrix} y^\dagger \\ z \end{bmatrix}, U_L X_L \begin{bmatrix} y^\dagger \\ z \end{bmatrix} > \\
&= \tfrac{1}{2} < U_L^\dagger U_L \begin{bmatrix} y^\dagger \\ z \end{bmatrix}, X_L \begin{bmatrix} y^\dagger \\ z \end{bmatrix} > \qquad (7.100) \\
&= \tfrac{1}{2} < \begin{bmatrix} y^\dagger \\ z \end{bmatrix}, X_L \begin{bmatrix} y^\dagger \\ z \end{bmatrix} > \\
&= <x,y,z>,
\end{aligned}
$$

since $U_L^\dagger = U_L^{-1}$.

Using 7.96 we may expand the trilinear form explicitly as follows:

$$
\boxed{
\begin{aligned}
<x,y,z> &= x^0y^0z^0 - \textstyle\sum_1^7 x^0y^az^a - \sum_1^7 x^ay^0z^a - \sum_1^7 x^ay^az^0 \\
&\quad + \textstyle\sum_1^7 x^{\{a}y^{a+1}z^{a+5\}} - \sum_1^7 x^{\{a}y^{a+5}z^{a+1\}},
\end{aligned}
} \qquad (7.101)
$$

where $\{...\}$ indicates sum over cyclic permutations of indices in the last two terms, which as usual are taken from 1 to 7, modulo 7. It is clear from 7.101 that $<x,y,z>$ is invariant with respect to permutations of $\{x,y,z\}$. That is, the vector and two spinor representations of $Spin(8)$ are interchangeable.

We can explicitly construct these representations as operators in $\mathbf{O}_L$ and $\mathbf{O}_R$. In particular, operation of the $so(8)$ basis elements listed in 7.97 (with an extra factor $\frac{1}{2}$ thrown in) on the spinor $\begin{bmatrix} y^\dagger \\ z \end{bmatrix}$ yields the simultaneous actions,

$$
z \rightarrow \frac{1}{2} e_{La}[z] \text{ and } y^\dagger \rightarrow -\frac{1}{2} e_{La}[y^\dagger],
$$

together with

$$
z \rightarrow \frac{1}{2} e_{Lab}[z] \text{ and } y^\dagger \rightarrow \frac{1}{2} e_{Lab}[y^\dagger].
$$

However,

$$-e_{La}[y^\dagger] = -e_{La}T_0[y] = T_0(e_{Ra}[y])$$

(see 7.19), and

$$e_{Lab}[y^\dagger] = e_{Lab}T_0[y] = T_0(-e_{Rab}[y]).$$

The corresponding actions on x are a little more difficult to extract. We start by taking the commutator of the generators 7.97 with the Clifford algebra vector X_L given in 7.95. This implies in the first case,

$$\begin{aligned}
x_L &\rightarrow -\tfrac{1}{2}e_{La}x_L - \tfrac{1}{2}x_L e_{La} \\
&= -\tfrac{1}{2}\textstyle\sum_{c=0}^{7}(e_{La}x^c e_{Lc} + x^c e_{Lc}e_{La}) \\
&= -\tfrac{1}{2}\textstyle\sum_{c=0}^{7}(e_{Lac} + e_{Lca})x^c \\
&= -(x^0 e_{La} - x^a e_{L0}) \\
&= -(x^0 e_a - x^a e_0)_L \\
&= (-\tfrac{1}{2}(e_{La} + e_{Ra})[X])_L \\
&= (-(e_{La}P_0 + P_0 e_{La})[x])_L \\
&= (-(e_{Ra}P_0 + P_0 e_{Ra})[x])_L,
\end{aligned}$$

where the last line is a result of 7.37,38, and is added to emphasize the left/right arbitrariness of this action on x (note that the action on z was left, and on y was right). The second action on x comes from

$$\begin{aligned}
x_L &\to \tfrac{1}{2}e_{Lab}x_L - \tfrac{1}{2}x_L e_{Lab} \\
&= \tfrac{1}{2}\textstyle\sum_{c=0}^{7}(e_{Lab}x^c e_{Lc} - x^c e_{Lc}e_{Lab}) \\
&= \tfrac{1}{2}\textstyle\sum_{c=0}^{7}(e_{Labc} - e_{Lcab})x^c \\
&= (x^a e_{Lb} - x^b e_{La}) \\
&= (x^a e_b - x^b e_a)_L \\
&= (\tfrac{1}{2}(e_{Lab} - (e_a e_b)_R)[X])_L \\
&= ((e_{La}P_0 e_{Lb} - e_{Lb}P_0 e_{La})[x])_L \\
&= ((e_{Ra}P_0 e_{Rb} - e_{Rb}P_0 e_{Ra})[x])_L.
\end{aligned}$$

These results are summarized below.

TRIALITY REPRESENTATIONS OF $so(8)$

$$\begin{array}{cc}
Spinor_1 & Spinor_1 \\
z \to \tfrac{1}{2}e_{La}[z] & z \to \tfrac{1}{2}e_{Lab}[z], \\
 & \\
Spinor_2 & Spinor_2 \\
y \to \tfrac{1}{2}e_{Ra}[y] & y \to -\tfrac{1}{2}e_{Rab}[y], \\
 & \\
Vector & Vector \\
x \to -(e_{La}P_0 + P_0 e_{La})[x] & x \to (e_{La}P_0 e_{Lb} - e_{Lb}P_0 e_{La})[x] \\
 & \\
= -\tfrac{1}{2}(e_{La} + e_{Ra}) & = \tfrac{1}{2}(e_{Lab} - (e_a e_b)_R)
\end{array} \tag{7.102}$$

It is interesting to note that the sum of those on the lefthand side is

$$\frac{1}{2}e_{La} + \frac{1}{2}e_{Ra} - \frac{1}{2}(e_{La} + e_{Ra}) = 0, \tag{7.103}$$

and on the righthand side is

$$\frac{1}{2}e_{Lab} - \frac{1}{2}e_{Rab} + \frac{1}{2}(e_{Lab} - e_{Rc}) = e_{Lab} - \frac{1}{2}(e_{Rab} + e_{Rc}), \tag{7.104}$$

where $e_a e_b = e_c$. My principal reason for mentioning this is that if I don't my equation numbers will be loused up for the rest of the chapter.

THE *Tri* IN TRIALITY

The map defined by

$$e_{La} \rightarrow e_{Ra},\ a = 1, ..., 7, \tag{7.105}$$

defines the triality map on the representations 7.102. That is, it cyclically permutes them. To see this we first complete the construction of the map 7.105:

$$e_{Lab} = e_{La}e_{Lb} \rightarrow e_{Ra}e_{Rb} = e_{Rba} = -e_{Rab}, \tag{7.106}$$

and

$$e_{Labc} = e_{La}e_{Lb}e_{Lc} \rightarrow e_{Ra}e_{Rb}e_{Rc} = e_{Rcba} = -e_{Rabc}. \tag{7.107}$$

[NOTE! It may appear from the derivation of the actions 7.106,107 that we have defined an algebraic automorphism. That this is not so, that in fact the map from $\mathbf{O}_L \rightarrow \mathbf{O}_R = \mathbf{O}_L$ is only a linear automorphism, can be seen by extending this construction one more step. Recall that

$$1_L = 1_R = e_{L7654321} = e_{R7654321} = e_{R1}e_{R2}e_{R3}e_{R4}e_{R5}e_{R6}e_{R7},$$

so that, for example,

$$e_{L123} = e_{L7654321}e_{L123} = -e_{L7654}.$$

According to 7.107, under the triality map

$$e_{L123} \rightarrow -e_{R123}.$$

However, assuming the map is a homomorphism we derive,

$$\begin{aligned} e_{L123} = -e_{L7}e_{L6}e_{L5}e_{L4} &\rightarrow -e_{R7}e_{R6}e_{R5}e_{R4} \\ &= -e_{R1}e_{R2}e_{R3}e_{R4}e_{R5}e_{R6}e_{R7}e_{R7}e_{R6}e_{R5}e_{R4} \\ &= -e_{R1}e_{R2}e_{R3} \\ &= e_{R3}e_{R2}e_{R1} \\ &= +e_{R123}, \end{aligned}$$

contradicting 7.107. Since every element of the adjoint octonions, whether a left or a right action, can be written as a linear combination of 1_L (transforming into itself), e_{La}, e_{Lab}, and e_{Labc}, the maps 7.105,106,107 completely determine a linear automorphism on $\mathbf{O}_L = \mathbf{O}_R$. Oddly, this map *does* preserve products (ie., acts as an algebra homomorphism) as long as the result of the product contains no elements of fourth or higher order!]

Denote the action of the triality map Tri. We observed above that

$$\frac{1}{2}e_{Ra} = (P_0 e_{La} + e_{La}P_0) - \frac{1}{2}e_{La}. \tag{7.108}$$

It is simply demonstrated that after taking the products and sums on the righthand side of 7.108, all surviving terms are of second order or less. Therefore,

$$Tri\{\frac{1}{2}e_{Ra}\} = Tri\{P_0\}Tri\{e_{La}\} + Tri\{e_{La}\}Tri\{P_0\} - Tri\{\frac{1}{2}e_{La}\}. \tag{7.109}$$

Observe that

$$\begin{aligned} Tri\{P_0\} &= Tri\{\tfrac{1}{8}(1 + e_{L476} + e_{L517} + e_{L621} + e_{L732} + e_{L143} + e_{L254} + e_{L365})\} \\ &= \tfrac{1}{8}(1 - e_{L476} - e_{L517} - e_{L621} - e_{L732} - e_{L143} - e_{L254} - e_{L365}) \\ &= \tfrac{1}{4} - P_0. \end{aligned} \tag{7.110}$$

Therefore,

$$\begin{aligned} Tri\{\tfrac{1}{2}e_{Ra}\} &= (\tfrac{1}{4} - P_0)e_{Ra} + e_{Ra}(\tfrac{1}{4} - P_0) - \tfrac{1}{2}e_{Ra} \\ &= -(P_0 e_{Ra} + e_{Ra}P_0) \\ &= -(P_0 e_{La} + e_{La}P_0) \\ &= -\tfrac{1}{2}(e_{La} + e_{Ra}). \end{aligned} \tag{7.111}$$

Likewise,

$$\begin{aligned} Tri\{-(P_0 e_{La} + e_{La}P_0)\} &= -((\tfrac{1}{4} - P_0)e_{Ra} + e_{Ra}(\tfrac{1}{4} - P_0)) \\ &= (P_0 e_{Ra} + e_{Ra}P_0) - \tfrac{1}{2}e_{Ra} \\ &= (P_0 e_{La} + e_{La}P_0) - \tfrac{1}{2}e_{Ra} \\ &= \tfrac{1}{2}e_{La}, \end{aligned} \tag{7.112}$$

by 7.108. In summary,

$$\boxed{\begin{aligned} Tri\{\tfrac{1}{2}e_{La}\} &= \tfrac{1}{2}e_{Ra} && \text{Spinor-Spinor} \\ Tri\{\tfrac{1}{2}e_{Ra}\} &= -(P_0 e_{La} + e_{La}P_0) && \text{Spinor-Vector} \\ &= -\tfrac{1}{2}(e_{La} + e_{Ra}) && \\ Tri\{-\tfrac{1}{2}(e_{La} + e_{Ra})\} &= \tfrac{1}{2}e_{La} && \text{Vector-Spinor.} \end{aligned}} \tag{7.113}$$

Demonstrating that Tri cyclically permutes the Lie algebra actions on the righthand side of 7.102 is a slightly more complicated problem. We already have that

$$Tri\{\frac{1}{2}e_{Lab}\} = -\frac{1}{2}e_{Rab}. \tag{7.114}$$

It should be clear that for all $a, b \in \{1, ..., 7\}$, $a \neq b$, such that $e_a e_b = e_c$,

$$\boxed{\begin{aligned} P_0 e_{Lab} &= P_0 e_{Lc}, \\ e_{Lab}P_0 &= e_{Lc}P_0, \\ P_0 e_{Lab}P_0 &= P_0 e_{Lc}P_0 = 0. \end{aligned}} \tag{7.115}$$

Therefore,

$$\begin{aligned} -\tfrac{1}{2}e_{Rab} &= \tfrac{1}{2}(2P_0-1)e_{Lab}(2P_0-1) \\ &= -(P_0e_{Lab}+e_{Lab}P_0)+\tfrac{1}{2}e_{Lab} \\ &= -(P_0e_{Lc}+e_{Lc}P_0)+\tfrac{1}{2}e_{Lab}. \end{aligned} \tag{7.116}$$

The reason for the last equality above is that the map Tri acts homomorphically on the last line, but *does not* on the penultimate line. That is, using the last equality above,

$$\begin{aligned} Tri\{-\tfrac{1}{2}e_{Rab}\} &= Tri\{-(P_0e_{Lc}+e_{Lc}P_0)+\tfrac{1}{2}e_{Lab}\} \\ &= -(\tfrac{1}{4}-P_0)e_{Rc}-e_{Rc}(\tfrac{1}{4}-P_0)-\tfrac{1}{2}e_{Rab} \\ &= -\tfrac{1}{2}e_{Rc}+(P_0e_{Rc}+e_{Rc}P_0)-\tfrac{1}{2}e_{Rab} \\ &= -\tfrac{1}{2}e_{Rc}+(P_0e_{Rc}+e_{Rc}P_0)-(P_0e_{Lc}+e_{Lc}P_0)+\tfrac{1}{2}e_{Lab} \\ &= \tfrac{1}{2}e_{Lab}-\tfrac{1}{2}e_{Rc}, \end{aligned} \tag{7.117}$$

since

$$(P_0e_{Rc}+e_{Rc}P_0)-(P_0e_{Lc}+e_{Lc}P_0)=(P_0e_{Lc}+e_{Lc}P_0)-(P_0e_{Lc}+e_{Lc}P_0)=0.$$

By 7.25,

$$\begin{aligned} -\tfrac{1}{2}e_{Rc} &= -\tfrac{1}{2}(e_ae_b)_R \\ &= -\tfrac{1}{2}((2e_{Lb}P_0e_{La}-2e_{La}P_0e_{Lb})+e_{La}e_{Lb}) \\ &= ((e_{La}P_0e_{Lb}-e_{Lb}P_0e_{La})-\tfrac{1}{2}e_{Lab}. \end{aligned} \tag{7.118}$$

Therefore,

$$Tri\{-\frac{1}{2}e_{Rab}\}=(e_{La}P_0e_{Lb}-e_{Lb}P_0e_{La}), \tag{7.119}$$

as expected.

Finally

$$
\begin{aligned}
Tri\{\tfrac{1}{2}e_{Lab} - \tfrac{1}{2}e_{Rc}\} &= -\tfrac{1}{2}e_{Rab} + e_{Rc}P_0 + P_0e_{Rc} \\
&= -\tfrac{1}{2}(e_{Rab} - 2e_{Rab}P_0 - 2P_0e_{Rab}) \\
&= -\tfrac{1}{2}(T_0e_{Rab}T_0) \\
&= \tfrac{1}{2}e_{Lab},
\end{aligned}
\qquad (7.120)
$$

completing the tricycle. In summary,

$$
\boxed{
\begin{array}{rcll}
Tri\{\tfrac{1}{2}e_{Lab}\} &=& -\tfrac{1}{2}e_{Rab} & \text{Spinor-Spinor} \\
Tri\{-\tfrac{1}{2}e_{Rab}\} &=& (e_{La}P_0e_{Lb} - e_{Lb}P_0e_{La}) & \\
&=& \tfrac{1}{2}(e_{Lab} - e_{Rc}) & \text{Spinor-Vector} \\
Tri\{\tfrac{1}{2}(e_{Lab} - e_{Rc})\} &=& \tfrac{1}{2}e_{Lab} & \text{Vector-Spinor}
\end{array}
}
\qquad (7.121)
$$

FREUDENTHAL'S PRINCIPLE OF TRIALITY

Let g_0, g_1, and g_2 be triality related elements of $so(8)$. That is,

$$Tri\{g_0\} = g_1, \quad Tri\{g_1\} = g_2, \quad Tri\{g_2\} = g_0.$$

The invariance 7.100 can then be written in the infinitesimal (Lie algebraic) form,

$$\boxed{< g_1[x], y, z > + < x, g_0[y], z > + < x, y, g_2[z] >= 0.} \qquad (7.122)$$

By 7.96 the second term can be expressed

$$
\begin{aligned}
< x, g_O[y], z > &= < (g_O[y])^\dagger, xz > && \text{inner product} \\
&= < T_O g_O[y], xz > \\
&= < T_O g_O T_O[y^\dagger], xz > && (7.123) \\
&= < y^\dagger, (T_O g_O T_O)^\dagger[xz] > \\
&= < y^\dagger, -T_O g_O T_O[xz] > .
\end{aligned}
$$

Together with 7.96, this implies we may rewrite 7.122

$$< y^\dagger, (g_1[x])z > + < y^\dagger, -T_O g_O T_O[xz] > + < y^\dagger, x(g_2[z]) >= 0,$$

which implies

$$\boxed{T_O g_O T_O[xz] = (g_1[x])z + x(g_2[z]),} \qquad (7.124)$$

which is an expression of Freudenthal's principle of triality [1]. We see from 7.123 that $so(8)$ almost acts as a derivation on the octonions (and of course, elements of LG_2, a Lie subalgebra of $so(8)$, do act as derivations, since G_2 is the automorphism group of **O**).

7.4. LG_2 and Tri.

In order to prove that some g in the adjoint algebra of **O** is an element of LG_2 we need merely prove it acts as a derivation on **O**. That is, that

$$g[xz] = (g[x])z + x(g[z]). \qquad (7.125)$$

Our claim is that LG_2 is spanned by elements of the form

$$g = e_{Lab} - e_{Lcd} : \ a, b, c, d \in \{1, ..., 7\} \text{ distinct}, \ g[1] = 0. \qquad (7.126)$$

Notice that in the particular case $a = 2,\ b = 3,\ c = 4,\ d = 6$,

$$
\begin{aligned}
Tri\{e_{L23} - e_{L46}\} &= -e_{R23} + e_{R46} \\
&= -\tfrac{1}{2}(e_{L7} - e_{L23} + e_{L46} + e_{L15}) \\
&\quad +\tfrac{1}{2}(e_{L7} + e_{L23} - e_{L46} + e_{L15}) \qquad (7.127) \\
&= e_{L23} - e_{L46}.
\end{aligned}
$$

Also,

$$\begin{aligned} T_0(e_{L23} - e_{L46})T_0 &= -e_{R23} + e_{R46} \\ &= e_{L23} - e_{L46}. \end{aligned} \tag{7.128}$$

As this same result can be duplicated for any g of the form 7.126, it should now be clear in general that for g of this form

$$Tri\{g\} = g,$$

and

$$T_0 g T_0 = g.$$

Combined with 7.124 this implies that

$$\begin{aligned} T_0 g T_0[xz] &= g[xz] \\ &= (Tri\{g\}[x])z + x(Tri^2\{g\}[z]) \\ &= (g[x])z + x(g[z]). \end{aligned}$$

Therefore g is a derivation of $\mathbf{O}$, and

$$g \in LG_2.$$

Since the set of such g is 14-dimensional, as is LG_2, the span of all such g is in fact LG_2, as claimed.

LG_2 TRIALITY TRIPLET

The copy of LG_2 spanned by elements of the form 7.126 generates the automorphism group of $\mathbf{O}$. Each element of this Lie algebra is invariant with respect to the action of the triality map Tri (and with respect to the map $g \to T_0 g T_0$). In this sense this copy of LG_2 is a triality singlet. There is also a triality triplet of copies of LG_2 which is constructed below.

The LG_2 defined by 7.126 can be seen to arise in the following way from the Clifford algebra $\mathbf{R}_{0,6} \simeq \mathbf{O}_L$. This can be represented in two natural ways

$(p, q, r \in \{1, ..., 6\})$:

$$\begin{array}{lcc} \text{scalar} & 1_L & 1_L \\ \text{vector} & e_{Lp} & e_{Lp7} \\ \text{2-vector} & e_{Lpq} & e_{Lpq} \\ \text{3-vector} & e_{Lpqr} & e_{Lpqr} \\ \text{4-vector} & e_{Lpq7} & e_{Lpq7} \\ \text{5-vector} & e_{Lp7} & e_{Lp} \\ \text{6-vector} & e_{L7} & e_{L7} \end{array} \tag{7.129}$$

Recall that under the commutator product the 1- and 2-vectors of $\mathbf{R}_{0,n}$ generate $so(n+1)$. Therefore, both the sets

$$\{e_{Lp}, e_{Lqr} : p, q, r \in \{1, ..., 6\}\} \tag{7.130}$$

and

$$\{e_{Lp7}, e_{Lqr} : p, q, r \in \{1, ..., 6\}\} = \{e_{Lab} : a, b \in \{1, ..., 7\}\} \tag{7.131}$$

form bases for $so(7)$. And *in both cases* the Lie subalgebras which kill the identity are copies of LG_2.

In the case of 7.131 this yields the LG_2 defined in 7.126. In the case 7.130 it yields a copy of LG_2 spanned by elements of the form

$$\boxed{g = e_{Lpq} - e_{Lrs} : \ p, q, r, s \in \{1, ..., 6\} \text{ distinct}, \ g[1] = 0,} \tag{7.132}$$

which generate $su(3)$, together with the 6-dimensional subspace of elements of the form (one for each $p = 1, ..., 6$)

$$\boxed{g = 2e_{Lp} - e_{Lqr} - e_{Lst} : \ p, q, r, s, t \in \{1, ..., 6\}, \ g[1] = 0.} \tag{7.133}$$

That yields $8 + 6 = 14$ elements all together. The 8-dimensional $su(3)$ subalgebra is shared by this copy of LG_2 and that defined in 7.126, so it is invariant under the action of Tri. The remaining 6-dimensional subspace is not shared, and is not Tri invariant. In particular, consider the action of the triality map on the element $2e_{L1} - e_{L26} - e_{L34}$ of this subspace.

$$\begin{aligned} Tri\{2e_{L1} - e_{L26} - e_{L34}\} &= 2e_{R1} + e_{R26} + e_{R34} \\ &= (-e_{L1} + e_{L26} + e_{L34} + e_{L57}) \\ &\quad + \tfrac{1}{2}(e_{L1} - e_{L26} + e_{L34} + e_{L57}) \\ &\quad + \tfrac{1}{2}(e_{L1} + e_{L26} - e_{L34} + e_{L57}) \\ &= 2e_{L57} + e_{L26} + e_{L34}. \end{aligned}$$

Therefore Tri maps the 6-dimensional space of elements 7.133 to the 6-dimensional space of elements of the form

$$\boxed{g = 2e_{Lp7} + e_{Lqr} + e_{Lst} : \ p, q, r, s, t \in \{1, ..., 6\}, \ g[e_7] = 0.} \tag{7.134}$$

These elements, together with the $su(3)$ in 7.132, close under the commutator product to form another copy of LG_2. In this case, however, the G_2 generated by these elements leaves e_7 invariant, instead of the identity.

Carrying on in this manner, we note that

$$\begin{aligned} Tri^2\{2e_{L1} - e_{L26} - e_{L34}\} &= Tri\{2e_{L57} + e_{L26} + e_{L34}\} \\ &= -2e_{R57} - e_{R26} - e_{R34} \\ &= -(e_{L1} + e_{L26} + e_{L34} - e_{L57}) \\ &\quad -\tfrac{1}{2}(+e_{L1} - e_{L26} + e_{L34} + e_{L57}) \\ &\quad -\tfrac{1}{2}(+e_{L1} + e_{L26} - e_{L34} + e_{L57}) \\ &= -2e_{L1} - e_{L26} - e_{L34}. \end{aligned} \tag{7.135}$$

Therefore Tri maps the 6-dimensional space of elements 7.134 to the 6-dimensional space of elements of the form

$$\boxed{g = -2e_{Lp} - e_{Lqr} - e_{Lst} : \ p, q, r, s, t \in \{1, ..., 6\}, \ g[e_7] = 0,} \tag{7.136}$$

which again with the $su(3)$ in 7.132 close under the commutator product to form another copy of LG_2.

Finally,

$$\begin{aligned} Tri^3\{2e_{L1} - e_{L26} - e_{L34}\} &= Tri\{-2e_{L1} - e_{L26} - e_{L34}\} \\ &= -2e_{R1} + e_{R26} + e_{R34} \\ &= -(-e_{L1} + e_{L26} + e_{L34} + e_{L57}) \\ &\quad +\tfrac{1}{2}(e_{L1} - e_{L26} + e_{L34} + e_{L57}) \\ &\quad +\tfrac{1}{2}(e_{L1} + e_{L26} - e_{L34} + e_{L57}) \\ &= 2e_{L1} - e_{L26} - e_{L34}, \end{aligned}$$

which is where we started from.

Therefore, the three LG_2's consisting of the $su(3)$ 7.132 together with the 6-dimensional spaces 7.133, or 7.134, or 7.136, form a triplet with respect to Tri. The first of these, like the LG_2 defined in 7.126, generate copies of G_2 that leave the identity invariant, hence act nontrivially on the 7-dimensional subspace of $\mathbf{O}$ spanned by $\{e_a, a = 1, ..., 7\}$. The last two generate G_2's that leave $e_7 \in \mathbf{O}$ invariant, hence act nontrivially on the 7-dimensional subspace of $\mathbf{O}$ spanned by $\{1, e_p, p = 1, ..., 6\}$.

It is interesting how often the (1,3)-motif arises in the context of the division algebras, but no speculation on the meaning of this example is offered at this point.

As noted earlier, if g takes the form 7.126, then

$$T_O g T_O = g. \tag{7.137}$$

Likewise, if g takes the form 7.132, then, for example,

$$\begin{aligned} T_O(2e_{L1} - e_{L26} - e_{L34})T_O &= -2e_{R1} + e_{R26} + e_{R34} \\ &= -(-e_{L1} + e_{L26} + e_{L34} + e_{L57}) \\ &\quad +\tfrac{1}{2}(e_{L1} - e_{L26} + e_{L34} + e_{L57}) \\ &\quad +\tfrac{1}{2}(e_{L1} + e_{L26} - e_{L34} + e_{L57}) \\ &= 2e_{L1} - e_{L26} - e_{L34}. \end{aligned}$$

That is, elements g of the form 7.132 also satisfy 7.137.

On the other hand,

$$\begin{aligned} T_O(2e_{L57} + e_{L26} + e_{L34})T_O &= -2e_{R57} - e_{R26} - e_{R34} \\ &= -(e_{L1} + e_{L26} + e_{L34} - e_{L57}) \\ &\quad -\tfrac{1}{2}(+e_{L1} - e_{L26} + e_{L34} + e_{L57}) \\ &\quad -\tfrac{1}{2}(+e_{L1} + e_{L26} - e_{L34} + e_{L57}) \\ &= -2e_{L1} - e_{L26} - e_{L34}. \end{aligned}$$

Therefore,

$$T_0(-2e_{L1} - e_{L26} - e_{L34})T_0 = T_0^2(2e_{L57} + e_{L26} + e_{L34})T_0^2$$
$$= 2e_{L57} + e_{L26} + e_{L34}.$$

So elements of the form 7.134 and 7.136 are interchanged by the map, $g \to T_0 g T_0$.

Therefore, for example, if g_0 is an element of the first LG_2 (7.132,133), g_1 the second (7.132,134), and g_2 the third (7.132,136), then by 7.124 (Freudenthal's principle of triality), for all $x, z \in \mathbf{O}$,

$$g_0[xz] = (g_1[x])z + x(g_2[z]),$$

since g_0 satisfies 7.137.

In the next section we'll give some meaning to this, and other, LG_2 triality triplets.

7.5. LG_2 Triplets and the X-Product.

Some preliminary results are needed. Let $g \in so(8)$, spanned by the set $\{e_{La},\ e_{Lab} :\ a, b \in \{1, ..., 7\}\}$. Note that

$$Tri\{T_o Tri\{e_{La}\}T_0\} = Tri\{T_0 e_{Ra} T_0\}$$
$$= Tri\{-e_{La}\}$$
$$= -e_{Ra}$$
$$= T_0 e_{La} T_0,$$

and

$$Tri\{T_o Tri\{e_{Lab}\}T_0\} = Tri\{T_0(-e_{Rab})T_0\}$$
$$= Tri\{e_{Lab}\}$$
$$= -e_{Rab}$$
$$= T_0 e_{Lab} T_0.$$

That is, in general,

$$\boxed{Tri\{T_0 Tri\{g\}T_0\} = T_0 g T_0,}$$
$$\boxed{Tri^2\{T_0 Tri^2\{g\}T_0\} = T_0 g T_0,} \tag{7.138}$$

where the second equality follows immediately from the first. These results, together with $Tri^3 = identity$, imply

$$\boxed{Tri\{T_0 g T_0\} = T_0 Tri^2\{g\}T_0,}$$
$$\boxed{Tri^2\{T_0 g T_0\} = T_0 Tri\{g\}T_0.} \tag{7.139}$$

As noted in section 7.2, if $A, B \in \mathbf{O}$, and X is a unit octonion (that is, $X \in S^7$; $XX^\dagger = X^\dagger X = 1$), then

$$\boxed{A \circ_X B = (AX)(X^\dagger B)} \tag{7.140}$$

defines a modified octonion product.

Let $g \in so(8)$, and note that $T_0(T_0 g T_0)T_0 = g$. This, together with 7.138,139, and multiple applications of 7.124, implies

$$\begin{aligned}
g[A \circ_X B] &= g[(AX)(X^\dagger B)] \\
&= (Tri\{T_0 g T_0\}[AX])(X^\dagger B) + (AX)(Tri^2\{T_0 g T_0\}[X^\dagger B]) \\
&= (T_0 Tri^2\{g\}T_0[AX])(X^\dagger B) + (AX)(T_0 Tri\{g\}T_0[X^\dagger B]) \\
&= (Tri^3\{g\}[A]X + ATri^4\{g\}[X])(X^\dagger B) \\
&\quad +(AX)(Tri^2\{g\}[X^\dagger]B + X^\dagger Tri^3\{g\}[B]) \\
&= (g[A]) \circ_X B + A \circ_X (g[B]) \\
&\quad +(ATri\{g\}[X])(X^\dagger B) + (AX)(Tri^2\{g\}[X^\dagger]B).
\end{aligned} \tag{7.141}$$

Let G_2^X be the automorphism group of the product $A \circ_X B$, and let LG_2^X be its Lie algebra. Then by 7.141,

$$\boxed{g \in LG_2^X \text{ iff } (ATri\{g\}[X])(X^\dagger B) + (AX)(Tri^2\{g\}[X^\dagger]B) = 0} \quad (7.142)$$

for all $A, B \in \mathbf{O}$.

Consider the special case of

$$X = \sum_1^7 X^a e_a \Longrightarrow X^\dagger = -X \Longrightarrow X \in S^6, \quad (7.143)$$

and in particular, and without a loss of generality, set

$$X = e_7. \quad (7.144)$$

Let

$$g \in e_{L7}e_{R7}LG_2 e_{R7}e_{L7}. \quad (7.145)$$

That is, there exists an $h \in LG_2$ such that

$$g = e_{L7}e_{R7}he_{R7}e_{L7} = e_{R7}e_{L7}he_{L7}e_{R7}. \quad (7.146)$$

Note that

$$h[1] = 0 \Longrightarrow g[1] = e_{L7}e_{R7}he_{R7}e_{L7}[1] = e_{R7}e_{L7}h[-1] = 0. \quad (7.147)$$

LG_2 can be divided into two parts: the 8-dimensional $su(3)$ spanned by

$$\{h = e_{Lpq} - e_{Lrs} : \ p,q,r,s \in \{1,...,6\} \text{ distinct }, \ h[1] = 0\}; \quad (7.148)$$

and the six dimensional subspace spanned by

$$\{h = 2e_{Lp7} - e_{Lqr} - e_{Lst} : \ p,q,r,s,t \in \{1,...,6\} \text{ distinct }, \ h[1] = 0\}. \quad (7.149)$$

If $h \in su(3)$, then $Tri\{h\} = Tri^2\{h\} = h$, and $h[e_7] = 0$. In addition,

$$h \in su(3) \Longrightarrow e_{L7}e_{R7}he_{R7}e_{L7} = h. \quad (7.150)$$

Therefore, 7.142 is trivially satisfied as

$$Tri\{g\}[e_7] = h[e_7] = 0 = h[-e_7] = Tri^2\{g\}[e_7^\dagger]. \quad (7.151)$$

Now suppose h is an element of the span of the set 7.149. In particular, and again without a loss of generality, set

$$h = 2e_{L57} - e_{L26} - e_{L34}. \tag{7.152}$$

With the help of multiple applications of the equations

$$\boxed{\begin{array}{l} e_{R1} = \frac{1}{2}(-e_{L1} + e_{L57} + e_{L26} + e_{L34}), \\ e_{R57} = \frac{1}{2}(e_{L1} - e_{L57} + e_{L26} + e_{L34}), \\ e_{R26} = \frac{1}{2}(e_{L1} + e_{L57} - e_{L26} + e_{L34}), \\ e_{R34} = \frac{1}{2}(e_{L1} + e_{L57} + e_{L26} - e_{L34}), \end{array}} \tag{7.153}$$

which are also valid with the subscripts R and L interchanged, we derive

$$\begin{aligned} g &= e_{R7}e_{L7}(2e_{L57} - e_{L26} - e_{L34})e_{L7}e_{R7} \\ &= e_{R7}(2e_{L57} + e_{L26} + e_{L34})e_{R7} \\ &= e_{R7}(2e_{R1} + e_{R26} + e_{R34})e_{R7} \\ &= 2e_{R1} - e_{R26} - e_{R34} \\ &= -2e_{L1} + e_{L26} + e_{L34}. \end{aligned} \tag{7.154}$$

Therefore,

$$\begin{aligned} Tri\{g\} &= -2e_{R1} - e_{R26} - e_{R34} \\ &= -2e_{L57} - e_{L26} - e_{L34} \\ &= -e_{L7}he_{L7}, \end{aligned} \tag{7.155}$$

and

$$\begin{aligned} Tri^2\{g\} &= 2e_{R57} + e_{R26} + e_{R34} \\ &= T_0gT_0 \\ &= -e_{R7}he_{R7}, \end{aligned} \tag{7.156}$$

since $T_0 h T_0 = h$ for all $h \in LG_2$. This implies

$$\begin{aligned} Tri\{g\}[e_7] &= -e_{L7} h e_{L7}[e_7] \\ &= e_{L7} h[1] \\ &= 0, \end{aligned} \tag{7.157}$$

and

$$\begin{aligned} Tri^2\{g\}[e_7] &= -e_{R7} h e_{R7}[e_7] \\ &= e_{R7} h[1] \\ &= 0. \end{aligned} \tag{7.158}$$

Therefore,

$$g \in LG_2^{e_7}.$$

Finally, since $-e_{L7} su(3) e_{L7} = -e_{R7} su(3) e_{R7} = su(3)$, every element of the 14-dimensional $e_{L7} e_{R7} LG_2 e_{R7} e_{L7}$ is in $LG_2^{e_7}$. In summary, on dimensional grounds we conclude that

$$\boxed{LG_2^{e_7} = e_{L7} e_{R7} LG_2 e_{R7} e_{L7},} \tag{7.159}$$

$$\boxed{Tri\{LG_2^{e_7}\} = -e_{L7} LG_2 e_{L7},} \tag{7.160}$$

$$\boxed{Tri^2\{LG_2^{e_7}\} = -e_{R7} LG_2 e_{R7}.} \tag{7.161}$$

This LG_2-triality triplet is the same as the one developed in the previous section. However, we can now generalize this result to all of S^6. That is,

$$\boxed{X = -X^\dagger \in S^6 \Longrightarrow LG_2^X = X_L X_R LG_2 X_R X_L.} \tag{7.162}$$

LG_2^X GENERAL SOLUTION

Now assume

$$X = X^0 + \sum_1^7 X^a e_a;\ X^0 \neq 0,\ \sum_1^7 X^a e_a \neq 0.$$

In particular, and again without a loss of generality, we may set

$$X = X^0 + X^7 e_7;\ X^0 \neq 0,\ X^7 \neq 0. \tag{7.163}$$

Let $g \in LG_2^X \subset so(8)$, and let

$$Z = Tri\{g\}[X]. \tag{7.164}$$

Since any infinitesimal $so(8)$ rotation of $X \in S^7$ is orthogonal to X (see section 7.2),

$$< Z, X >= \frac{1}{2}(Z^\dagger X + X^\dagger Z) = \frac{1}{2}(Z X^\dagger + X Z^\dagger) = 0. \tag{7.165}$$

Therefore, Z takes the form

$$\begin{aligned} Z &= m e_7 X + nU, \qquad m, n \in \mathbf{R},\ U = \textstyle\sum_1^6 U^p e_p \in S^6, \\ Z^\dagger &= -m e_7 X^\dagger - nU. \end{aligned} \tag{7.166}$$

$g \in LG_2^X$ implies

$$g[(AX)(X^\dagger B)] = (g[A]X)(X^\dagger B) + (AX)(X^\dagger g[B]) \tag{7.167}$$

for all $A, B \in \mathbf{O}$. Setting $A = B = 1$ implies

$$g[1] = g[1] + g[1],$$

so

$$\boxed{g \in LG_2^X \Longrightarrow g[1] = 0.} \tag{7.168}$$

$g[1] = 0$ implies g is an element of the vector $so(7)$ spanned by

$$\{e_{La} - e_{Lbc} :\ (e_{La} - e_{Lbc})[1] = 0\}. \tag{7.169}$$

Consider the $so(7)$ element $e_{L7} - e_{L23}$, which satisfies

$$\begin{aligned} Tri\{e_{L7} - e_{L23}\} &= e_{R7} + e_{R23} \\ &= e_{L15} + e_{L46}. \end{aligned}$$

Therefore, the elements of this $Tri\{so(7)\}$ are linear in the $\mathbf{O}_L$ bilinears e_{Lab}, which satisfy

$$Tri\{e_{Lab}\} = -e_{Rab} = T_0 e_{Lab} T_0.$$

So

$$\boxed{g \in LG_2^X \Longrightarrow Tri^2\{g\} = T_0 Tri\{g\} T_0.} \tag{7.170}$$

By 7.139 this implies

$$Tri^2\{g\} = Tri^2\{T_0 g T_0\}.$$

Applying Tri to both sides implies

$$\boxed{g \in LG_2^X \Longrightarrow g = T_0 g T_0.} \tag{7.171}$$

Again setting $Z = Tri\{g\}[X]$, we find

$$\begin{aligned} Tri^2\{g\}[X^\dagger] &= T_0 T_0 Tri^2\{g\} T_0[X] \\ &= T_0 Tri\{g\}[X] \\ &= T_0[Z] \\ &= Z^\dagger. \end{aligned} \tag{7.172}$$

Therefore, 7.142 may be written in the form

$$(AZ)(X^\dagger B) + (AX)(Z^\dagger B) = 0. \tag{7.173}$$

Given 7.166 this may be written

$$\begin{aligned} H(A,B) &= m(Ae_7X)(X^\dagger B) - m(AX)(e_7X^\dagger B) \\ &\quad + n(AU)(X^\dagger B) - n(AX)(UB) \\ &= 0. \end{aligned} \tag{7.174}$$

So

$$H(A,X) = m(Ae_7X) - m(AX)e_7 + n(AU) - n(AX)(UX) = 0.$$

But $m(Ae_7X) - m(AX)e_7 = 0$, since $X = X^0 + X^7e_7$, so

$$H(A,X) = n(AU) - n(AX)(UX) = n(AU) - n(AX)(X^\dagger U) = 0. \quad (7.175)$$

Since U is orthogonal to X and $X^\dagger$, the only way 7.175 can hold for all $A \in \mathbf{O}$ is if $n = 0$. Therefore,

$$H(A,B) = m(Ae_7X)(X^\dagger B) - m(AX)(e_7X^\dagger B) = 0.$$

But

$$H(e_1X^\dagger, Xe_2) = m(e_1e_7)e_2 - me_1(e_7e_2) = -2me_4 = 0,$$

so finally, $m = 0$ too, and in general

$$\boxed{g \in LG_2^X \text{ iff } Tri\{g\}[X] = Tri^2\{g\}[X^\dagger] = 0,} \quad (7.176)$$

so each of the terms in 7.142 is separately zero (given 7.142, the reverse implication is trivial).

Now we are ready to derive more central results. Application of the results above, together with 7.124, imply

$$\begin{aligned} g[AX^\dagger] &= T_0gT_0[AX^\dagger] \\ &= (Tri\{g\}[A])X^\dagger + A(Tri^2\{g\}[X^\dagger]) \\ &= (Tri\{g\}[A])X^\dagger. \end{aligned}$$

In addition,

$$\begin{aligned} g[AX^\dagger] &= g[((AX^\dagger)X)(X^\dagger(1))] \\ &= ((g[AX^\dagger]X)X^\dagger + ((AX^\dagger)X)(X^\dagger g[1]) \\ &= (X_RgX_R^\dagger[A])X^\dagger, \end{aligned}$$

since $g[1] = 0$. Therefore,

$$\boxed{g \in LG_2^X \Longrightarrow Tri\{g\} = X_RgX_R^\dagger.} \quad (7.177)$$

Together with 7.170,171, this implies

$$\boxed{g \in LG_2^X \Longrightarrow Tri^2\{g\} = X_L^\dagger g X_L.} \tag{7.178}$$

These results are entirely consistent with 7.159,160,161, but they do not actually solve for LG_2^X in this general case the way that 7.159 does in the case $X = e_7$.

Note that with $X = X^0 + X^7 e_7$, and identifying

$$su(3) = \{g \in LG_2 : g[e_7] = 0\} \tag{7.179}$$

(as usual, the Lie algebra of the stability group of e_7), then

$$\begin{aligned} g[(AX)(X^\dagger B)] &= (g[A]X)(X^\dagger B) + (Ag[X])(X^\dagger B) + (AX)(g[X^\dagger]B) + (AX)(X^\dagger g[B]) \\ &= (g[A]X)(X^\dagger B) + (AX)(X^\dagger g[B]), \end{aligned} \tag{7.180}$$

since

$$g[X] = X^0 g[1] + X^7 g[e_7] = 0. \tag{7.181}$$

That is,

$$\boxed{su(3) \subset LG_2^X,} \tag{7.182}$$

accounting for 8 of the 14 dimensions of LG_2^X.

The other 6 dimensions of LG_2 (not LG_2^X) are found in

$$LG_2/su(3) =$$

$$\text{span}\{h = 2e_{Lp7} - e_{Lqr} - e_{Lst} : p,q,r,s,t \in \{1,...,6\} \text{ distinct }, h[1] = 0\}. \tag{7.183}$$

My claim is (and I will spare you the tedious details of its derivation)

$$\boxed{\begin{aligned} &LG_2^X/su(3) = \\ &\text{span}\{g = 2X_L e_{Lp7} - X_R^\dagger(e_{Lqr} + e_{Lst}) : 2e_{Lp7} - e_{Lqr} - e_{Lst} \in LG_2/su(3)\}. \end{aligned}} \tag{7.184}$$

We investigate this claim with the element (the different forms of which are derived using 7.153 and its cyclic variations)

$$\boxed{\begin{aligned} g &= 2X_L e_{L57} - X_R^\dagger(e_{L26} + e_{L34}) \\ &= 2X_L e_{L57} - X_R^\dagger(e_{R1} + e_{R57}) \\ &= X^0(2e_{L57} - e_{L26} - e_{L34}) \\ &\quad +X^7(2e_{L5} - e_{L63} - e_{L24}) \\ &= X^0(-2e_{R57} + e_{R26} + e_{R34}) \\ &\quad +X^7(-2e_{R5} + e_{R63} + e_{R24}) \\ &= -2X_R^\dagger e_{R57} + X_L(e_{L1} + e_{L57}) \\ &= \tfrac{1}{2}X_L(3e_{L57} + e_{L1}) - \tfrac{1}{2}X_R^\dagger(3e_{R57} + e_{R1}). \end{aligned}} \tag{7.185}$$

The last equality above makes it clear that

$$g = T_0 g T_0. \tag{7.186}$$

From this equality we also derive

$$\begin{aligned} g[1] &= \tfrac{1}{2}X_L(3e_{L57} + e_{L1})[1] - \tfrac{1}{2}X_R^\dagger(3e_{R57} + e_{R1})[1] \\ &= X_L[2e_1] - X_R^\dagger[2e_1] \\ &= 2Xe_1 - 2e_1X^\dagger \\ &= 0, \end{aligned} \tag{7.187}$$

since $Xe_p = e_pX^\dagger$, $p = 1, ..., 6$. This verifies 7.171 and 7.168.

Using the third equality in 7.185 we determine (again using 7.153)

$$\begin{aligned} Tri\{g\} &= X^0(-2e_{R57} + e_{R26} + e_{R34}) \\ &\quad + X^7(2e_{R5} + e_{R63} + e_{R24}) \\ &= -2(X^0 e_{R57} + X^7 e_{R7} e_{R57}) + X^0(e_{L1} + e_{L57}) + X^7(e_{L5} + e_{L71}) \\ &= -2X_R e_{R57} + X_L(e_{L1} + e_{L57}) \\ &= -2e_{R57} X_R^\dagger + (e_{L1} + e_{L57}) X_L^\dagger. \end{aligned} \tag{7.188}$$

Therefore,

$$\begin{aligned} Tri\{g\}[X] &= -2e_{R57} X_R^\dagger[X] + (e_{L1} + e_{L57}) X_L^\dagger[X] \\ &= -2e_{R57}[1] + (e_{L1} + e_{L57})[1] \\ &= -2e_1 + e_1 + e_1 \\ &= 0, \end{aligned}$$

which verifies half of 7.176. As to the other half, using 7.153 rewrite

$$\begin{aligned} Tri\{g\} &= X^0(2e_{L57} - e_{L26} - e_{L34}) \\ &\quad + X^7(2e_{L71} + e_{L63} + e_{L24}), \end{aligned}$$

so that

$$\begin{aligned} Tri^2\{g\} &= X^0(-2e_{R57} + e_{R26} + e_{R34}) \\ &\quad + X^7(-2e_{R71} - e_{R63} - e_{R24}) \\ &= X^0(2e_{L57} - e_{L26} - e_{L34}) \\ &\quad + X^7(-2e_{L5} - e_{L63} - e_{L24}) \\ &= T_0 Tri\{g\} T_0 \\ &= 2e_{L57} X_L - (e_{R1} + e_{R57}) X_R. \end{aligned} \tag{7.189}$$

Therefore,

$$\begin{aligned} Tri^2\{g\}[X^\dagger] &= Tri^2\{g\}T_0[X] \\ &= T_0 Tri\{g\}[X] \\ &= 0. \end{aligned}$$

Finally, in satisfying all of 7.176, this proves

$$g \in LG_2^X,$$

and proves 7.184 in general. I leave it to the reader to prove 7.177 and 7.178. Note, for example, that 7.177 may be rewritten

$$g = X_R^\dagger Tri\{g\} X_R,$$

and that

$$\begin{aligned} X_R^\dagger(e_{L1} + e_{L57})X_R &= X_R^\dagger(e_{R26} + e_{R34})X_R \\ &= (e_{R26} + e_{R34}) \\ &= (e_{L1} + e_{L57}). \end{aligned}$$

Finally note that if $X^0 = 1$, and $X^7 = 0$, then by 7.185

$$g = 2e_{L57} - e_{L26} - e_{L34} \in LG_2.$$

And if $X^0 = 0$, and $X^7 = 1$, then

$$\begin{aligned} g &= 2e_{L5} - e_{L63} - e_{L24} \\ &= e_{L7}(2e_{L5} + e_{L63} + e_{L24})e_{L7} \\ &= e_{L7}(2e_{R5} + e_{R63} + e_{R24})e_{L7} \\ &= e_{L7}e_{R7}(2e_{R71} - e_{R63} - e_{R24})e_{R7}e_{L7}, \end{aligned}$$

where

$$2e_{R71} - e_{R63} - e_{R24} \in LG_2,$$

which verifies that

$$LG_2^{e_7} = e_{L7}e_{R7}LG_2e_{R7}e_{L7}.$$

LG_2^X AND THE X-ADJOINT ALGEBRA $\mathbf{O}_{LX}$

Let $\mathbf{O}_X$ be the X-product modified octonions, and let e_{aX}, $a = 1, ..., 7$, be a basis for the pure hypercomplex subspace of $\mathbf{O}_X$, arranged so that

$$\boxed{e_{aX} \circ_X e_{(a+1)X} = e_{(a+5)X}.} \tag{7.190}$$

For simplicity again set

$$X = X^0 + X^7 e_7. \tag{7.191}$$

In this case if we set

$$\boxed{\begin{aligned} e_{mX} &= Xe_m,\ m = 1,2,3,5, \\ e_{kX} &= e_k,\ k = 4,6,7, \end{aligned}} \tag{7.192}$$

then 7.188 is satisfied.

To see this let

$$\begin{aligned} J_+ &= \tfrac{1}{2}(1 + I_1) = P_0 + P_4 + P_6 + P_7, \\ J_- &= \tfrac{1}{2}(1 - I_1) = P_1 + P_2 + P_3 + P_5, \end{aligned} \tag{7.193}$$

and note that for all $A \in \mathbf{O}$, 7.192 is equivalent to

$$\boxed{A_X = J_+[A] + XJ_-[A] = J_+[A] + J_-[A]X^\dagger.} \tag{7.194}$$

Then 7.190 is equivalent to

$$\boxed{A_X \circ_X B_X = (AB)_X} \tag{7.195}$$

for all $A, B \in \mathbf{O}$.

It is easily proven that

$$\begin{aligned} J_+[AB] &= J_+[A]J_+[B] + J_-[A]J_-[B], \\ J_-[AB] &= J_+[A]J_-[B] + J_-[A]J_+[B], \end{aligned} \tag{7.196}$$

for all $A, B \in \mathbf{O}$. Also,

$$\begin{aligned} (J_+[A]X)(X^\dagger J_+[B]) &= J_+[A]J_+[B], \\ XJ_-[A]X = J_-[A] &= X^\dagger J_-[A]X^\dagger. \end{aligned} \tag{7.197}$$

By 7.74

$$\begin{aligned}(J_+[A]X)J_-[B] &= (J_+[A]X)(X^\dagger(XJ_-[B])) \\ &= (J_+[A](XJ_-[B]X))X^\dagger \\ &= (J_+[A]J_-[B])X^\dagger \\ &= X(J_+[A]J_-[B]),\end{aligned} \tag{7.198}$$

since $J_+[A]J_-[B]$ is linear in e_m, $m = 1, 2, 3, 5$.

Taking the Hermitian conjugate of 7.74, and interchanging the indices a and b, implies

$$\boxed{(e_a X)(X^\dagger e_b) = X((X^\dagger e_a)e_b).} \tag{7.199}$$

Therefore,

$$\begin{aligned}J_-[A](X^\dagger J_+[B]) &= ((J_-[A]X^\dagger)X)(X^\dagger J_+[B]) \\ &= X((X^\dagger J_-[A]X^\dagger)J_+[B]) \\ &= X(J_-[A]J_+[B]).\end{aligned} \tag{7.200}$$

Combining 7.196-200 implies

$$\begin{aligned}A_X \circ_X B_X &= ((J_+[A] + J_-[A]X^\dagger)X)(X^\dagger(J_+[B] + XJ_-[B])) \\ &= (J_+[A]X + J_-[A])(X^\dagger J_+[B] + J_-[B]) \\ &= (J_+[A]J_+[B] + J_-[A]J_-[B]) + ((J_+[A]X)J_-[B] + J_-[A](X^\dagger J_+[B \\ &= J_+[AB] + X(J_+[A]J_-[B] + J_-[A]J_+[B]) \\ &= J_+[AB] + XJ_-[AB] \\ &= (AB)_X,\end{aligned} \tag{7.201}$$

thereby proving 7.195.

Let $\mathbf{O}_{LX}$ be the associative left adjoint algebra of $\mathbf{O}_X$. Let $A, B \in \mathbf{O}$, and $A_X, B_X \in \mathbf{O}_X$ their images via the isomorphism 7.194. Then we define

$$\boxed{A_{LX}[B_X] = A_X \circ_X B_X.} \tag{7.202}$$

Therefore, for $k = 4, 6, 7$,

$$\begin{aligned} e_{LkX}[B_X] &= (e_kX)(X^\dagger B_X) \\ &= (e_k(B_XX))X^\dagger \\ &= X_R^\dagger e_{Lk}X_R[B_X]. \end{aligned} \tag{7.203}$$

But if $k = 7$, then

$$e_{LkX} = X_R^\dagger e_{L7}X_R = e_{L7}. \tag{7.204}$$

Also, for $m = 1, 2, 3, 5$,

$$\begin{aligned} e_{LmX}[B_X] &= (e_{mX}X)(X^\dagger B_X) \\ &= (Xe_mX)(X^\dagger B_X) \\ &= e_m(X^\dagger B_X) \\ &= e_{Lm}X_L^\dagger[B_X] \\ &= X_Le_{Lm}[B_X]. \end{aligned} \tag{7.205}$$

Because of the $e_a \to e_{aX}$ isomorphism it is now clear that

$$\boxed{(e_{LaX}e_{LbX} - e_{LcX}e_{LdX}) \in LG_2^X \text{ iff } (e_{La}e_{Lb} - e_{Lc}e_{Ld}) \in LG_2.} \tag{7.206}$$

Consider the specific case $X = e_7$. The equations 7.203-205 imply in this case,

$$\begin{aligned} &e_{L1X} = e_{L71}, \quad e_{L2X} = e_{L72}, \quad e_{L3X} = -e_{L37}, \\ &e_{L4X} = e_{L67}, \quad e_{L5X} = -e_{L57}, \quad e_{L6X} = e_{L74}, \\ &e_{L7X} = e_{L7}. \end{aligned} \tag{7.207}$$

Let $SU^X(3) \subset G_2^X$ be the stability group of $e_{7X} = e_7$. A basis for its

8-dimensional Lie algebra consists of

$$
\begin{aligned}
e_{L2X}e_{L6X} - e_{L3X}e_{L4X} &= e_{L24} - e_{L63} \in su(3),\\
e_{L6X}e_{L1X} - e_{L4X}e_{L5X} &= e_{L41} - e_{L56} \in su(3),\\
e_{L4X}e_{L1X} - e_{L5X}e_{L6X} &= -e_{L61} + e_{L45} \in su(3),\\
e_{L5X}e_{L2X} - e_{L1X}e_{L3X} &= e_{L52} - e_{L13} \in su(3),\\
e_{L6X}e_{L3X} - e_{L2X}e_{L4X} &= -e_{L34} + e_{L26} \in su(3),\\
e_{L1X}e_{L2X} - e_{L3X}e_{L5X} &= e_{L12} - e_{L35} \in su(3),\\
e_{L2X}e_{L3X} - e_{L4X}e_{L6X} &= e_{L23} - e_{L46} \in su(3),\\
e_{L2X}e_{L3X} - e_{L1X}e_{L5X} &= e_{L23} - e_{L15} \in su(3).
\end{aligned}
\qquad (7.208)
$$

That is, $su^{e_7}(3) = su(3)$, as expected ($su(3) = su^1(3)$).

As to the rest of $LG_2^{e_7}$, consider the element

$$2e_{L5X}e_{L7X} - e_{L2X}e_{L6X} - e_{L3X}e_{L4X} = 2e_{L5} - e_{L24} - e_{L63}, \qquad (7.209)$$

which is what you get from 7.185 if you set $X^0 = 0$, and $X^7 = 1$. Similar results hold for the other elements of $LG_2^{e_7}/su(3)$, which demonstrates the consistency of the O_{LX} approach with what went before.

Using $SU(3)$ we can construct a nicer looking isomorphism from $\mathbf{O}$ to $\mathbf{O}_X$. Redefine (where now $X \in S^7$ is arbitrary),

$$A_X = X^{1/3}AX^{-1/3}. \qquad (7.210)$$

(note that $X^{1/3}$ is not uniquely defined.) In this case,

$$\boxed{(X^{1/3}AX^{-1/3}) \circ_X (X^{1/3}BX^{-1/3}) = X^{1/3}(AB)X^{-1/3},} \qquad (7.211)$$

which gives S^7 an almost group structure (what I would call a sliding group structure; sliding on the X-product). This also makes it clear that if $X = \pm 1$ (that is, $X^{1/3}$ is a $6th$ root of unity), then the X-product is ordinary octonion multiplication, and $A \to X^{1/3}AX^{-1/3}$ is a G_2 action (see also [7],[8]).

Much more could be said, and I dare say it eventually will be.

8. Corridors.

8.1. Magic Square.

The Freudenthal-Tits Magic Square [15] is a 4×4 array of groups associated with the Jordan algebras of 3×3 Hermitian matrices over **R**, **C**, **Q**, or **O**, complexified by (tensored with) **R**, **C**, **Q**, or **O**. That is, each group is associated with a pair of division algebras.

The mathematical context upon which the model building of chapters 3 through 6 rests relied heavily on treating the (tensored) division algebras as spinor spaces of their left adjoint algebras (identified as Clifford algebras), of tensoring those adjoint algebras with $\mathbf{R}(2)$ (doubling the size of the spinor space), and of treating the parts of the right adjoint algebras that commute with the left as internal. These exact same methods will be employed here to generate bases for the Lie algebras of the groups of a version of the magic square. Each will be derived from a tensor product of two division algebras, and as the order in which they are taken is immaterial, only ten distinct Lie algebras will result.

The foundation upon which the method rests is $\mathbf{R}(2)$. In $\mathbf{R}(2)$ define again

$$\epsilon=\begin{bmatrix}1&0\\0&1\end{bmatrix},\alpha=\begin{bmatrix}1&0\\0&-1\end{bmatrix},$$

$$\beta=\begin{bmatrix}0&1\\1&0\end{bmatrix},\omega=\begin{bmatrix}0&1\\-1&0\end{bmatrix}. \qquad (8.1)$$

Let $\mathbf{K}\otimes\mathbf{K}'$ be the tensor product of two of the four division algebras. Let c_k and c'_l denote bases for the pure hypercomplex parts of **K** and **K**′ (ie., i, q_j, or e_a). In $\mathbf{K}_L\otimes\mathbf{K}'_L(2)$ the elements

$$\omega,\quad c_{Lk}\alpha,\quad c'_{Ll}\beta \qquad (8.2)$$

anticommute (and associate) and form the basis for the 1-vector generators of a (possibly nonuniversal, meaning the product of these 1-vectors may equal the identity) Clifford algebra with negative definite Euclidean metric. Under commutation they generate the 2-vectors

$$c_{Lk}\beta, \quad c'_{Ll}\alpha, \quad c_{Lk_1k_2}\epsilon, \quad c'_{Ll_1l_2}\epsilon, \quad c_{Lk}c'_{Ll}\omega. \tag{8.3}$$

Together the elements 7.2 and 7.3 form a basis for a representation of the Lie algebra $so(\dim\mathbf{K}+\dim\mathbf{K}')$. I'll denote this $External_{KK'}$, and call it the external subalgebra.

To this collection we now add the spinors of $\mathbf{K}_L \otimes \mathbf{K}'_L(2)$, namely the elements of $(\mathbf{K}\otimes\mathbf{K}')^2$, without yet specifying a commutator product on this linear space. I'll denote this $Spinor_{KK'}$.

Finally we add all the elements of $\mathbf{K}_R \otimes \mathbf{K}'_R$ whose action on the spinor space is distinct from and commutes with (ie., is internal with respect to) the action of $so(\dim\mathbf{K} + \dim\mathbf{K}')$ on the spinor space, and which together generate a compact subgroup. I'll denote this $Internal_{KK'}$, and call it the internal subalgebra. Note that because of the nonassociativity of $\mathbf{O}$, only $\mathbf{O}_R$ will be excluded by these conditions ($\mathbf{C}_R$ and $\mathbf{Q}_R$ both commute with $External_{KK'}$, whatever the algebras $\mathbf{K}$ and $\mathbf{K}'$). That $\mathbf{O}_R$ *can* be excluded is in a sense responsible for the existence of the exceptional Lie groups F_4, E_6, E_7, and E_8.

The total resulting linear space will be denoted $MS_{KK'}$, MS for magic square.

8.2. The Ten $MS_{KK'}$.

$\mathbf{R}\otimes\mathbf{R}$

$\mathbf{R}$ has no complex part, so the external subalgebra is spanned by ω (1-vector basis for $\mathbf{R}_{0,1}$).

$$External_{RR} = u(1) = so(2) \tag{8.4}$$

(there is no $Internal_{RR}$). $Spinor_{RR}$ is 2-dimensional, so MS_{RR} has a total of 3 dimensions and we make the identification:

$$MS_{RR} = so(3) = su(2). \tag{8.5}$$

$\mathbf{R} \otimes \mathbf{C}$

$External_{RC}$ is spanned by ω and $i\beta$ (1-vector basis for $\mathbf{R}_{0,2}$), and $i\alpha$ (2-vector).

$$External_{RC} = su(2) = so(3). \tag{8.6}$$

$Spinor_{RC}$ is 4-dimensional, and

$$Internal_{RC} = u(1) = so(2) \tag{8.7}$$

is spanned by i_R (ie., its action on $\begin{bmatrix} x \\ y \end{bmatrix} \in Spinor_{RC}$ is $i_R \begin{bmatrix} x \\ y \end{bmatrix} = \begin{bmatrix} x \\ y \end{bmatrix} i = i \begin{bmatrix} x \\ y \end{bmatrix}$). That's 8 elements altogether, and we make the identification:

$$MS_{RC} = su(3). \tag{8.8}$$

$\mathbf{R} \otimes \mathbf{Q}$

$External_{RQ}$ is spanned by ω and $q_{Lj}\beta$ (1-vector basis for $\mathbf{R}_{0,4}$), and $q_{Lj}\alpha$, $q_{Lj}\epsilon$ (2-vectors).

$$External_{RQ} = sp(2) = so(5). \tag{8.9}$$

$Spinor_{RQ}$ is 8-dimensional, and

$$Internal_{RQ} = sp(1) = so(3) \tag{8.10}$$

is spanned by q_{Rj} . That's 10+8+3 = 21 elements altogether, and we make the identification:

$$MS_{RQ} = sp(3). \tag{8.11}$$

$\mathbf{R} \otimes \mathbf{O}$

$External_{RO}$ is spanned by ω and $e_{La}\beta$ (1-vector basis for $\mathbf{R}_{0,8}$), and $e_{La}\alpha$, $e_{Lab}\epsilon$ (2-vectors).

$$External_{RO} = so(9). \tag{8.12}$$

$Spinor_{RO}$ is 16-dimensional, but because $\mathbf{O}_L = \mathbf{O}_R$, $Internal_{RO}$ is empty. (The inclusion of $\mathbf{O}_R$ causes $so(9)$ to expand to $so(16)$, leading to a quite different result.) That's 36+16 = 52 elements altogether, and we make the identification:

$$MS_{RO} = LF_4, \tag{8.13}$$

where LF_4 is the Lie algebra of F_4.

$\mathbf{C} \otimes \mathbf{C}$

Let i and i' be distinct imaginary units that commute with each other. $External_{RC}$ is spanned by ω, $i\alpha$, $i'\beta$ (1-vector basis for nonuniversal $\mathbf{R}_{0,3}$), and $i\beta$, $i'\alpha$, $ii'\omega$ (2-vector).

$$External_{CC} = so(4) = su(2) \times su(2). \tag{8.14}$$

$Spinor_{CC}$ is 8-dimensional, and

$$Internal_{CC} = u(1) \times u(1) \tag{8.15}$$

is spanned by i_R and i'_R . That's 6+8+2 =16 elements altogether, and we make the identification:

$$MS_{CC} = su(3) \times su(3). \tag{8.16}$$

$\mathbf{C} \otimes \mathbf{Q}$

$External_{CQ}$ is spanned by ω, $i\alpha$, $q_{Lj}\beta$ (1-vector basis for $\mathbf{R}_{0,5}$), and $i\beta$, $q_{Lj}\alpha$, $iq_{Lj}\omega$, $q_{Lj}\epsilon$ (2-vectors).

$$External_{CQ} = so(6). \tag{8.17}$$

$Spinor_{CQ}$ is 16-dimensional, and

$$Internal_{CQ} = u(1) \times su(2) \tag{8.18}$$

is spanned by i and q_{Rj} . That's 15+16+4 = 35 elements altogether, and we make the identification:

$$MS_{CQ} = su(6). \tag{8.19}$$

$\mathbf{C} \otimes \mathbf{O}$

$External_{CO}$ is spanned by ω, $i\alpha$, $e_{La}\beta$ (1-vector basis for $\mathbf{R}_{0,9}$), and $i\beta$, $e_{La}\alpha$, $ie_{La}\omega$, $e_{Lab}\epsilon$ (2-vectors).

$$External_{CO} = so(10). \tag{8.20}$$

$Spinor_{CO}$ is 32-dimensional, and again because $\mathbf{O}_L = \mathbf{O}_R$, $Internal_{RO}$ is spanned only by i_R.

$$Internal_{CO} = u(1) = so(2). \tag{8.21}$$

That's 45+32+1 = 78 elements altogether, and we make the identification:

$$MS_{CO} = LE_6. \tag{8.22}$$

$\mathbf{Q} \otimes \mathbf{Q}$

Let q_j and q'_j be distinct and mutually commuting bases for the hypercomplex quaternions. $External_{QQ}$ is spanned by ω, $q_{Lj}\alpha$, $q'_{Lj}\beta$ (1-vector basis for nonuniversal $\mathbf{R}_{0,7}$), and $q_{Lj}\beta$, $q'_{Lj}\alpha$, $q_{Lj}q'_{Lk}\omega$, $q_{Lj}\epsilon$, $q'_{Lj}\epsilon$ (2-vectors).

$$External_{QQ} = so(8). \tag{8.23}$$

$Spinor_{QQ}$ is 32-dimensional, and

$$Internal_{QQ} = su(2) \times su(2) \tag{8.24}$$

is spanned by q_{Rj} and q'_{Rj} . That's 28+32+6 = 66 elements altogether, and we make the identification:

$$MS_{QQ} = so(12). \tag{8.25}$$

$\mathbf{Q} \otimes \mathbf{O}$

$External_{QO}$ is spanned by ω, $q_{Lj}\alpha$, $e_{La}\beta$ (1-vector basis for nonuniversal $\mathbf{R}_{0,11}$), and $q_{Lj}\beta$, $e_{La}\alpha$, $q_{Lj}e_{La}\omega$, $q_{Lj}\epsilon$, $e_{Lab}\epsilon$ (2-vectors).

$$External_{QO} = so(12). \tag{8.26}$$

$Spinor_{QO}$ is 64-dimensional, and again because $\mathbf{O}_L = \mathbf{O}_R$, $Internal_{RO}$ is spanned only by q_{Rj}.

$$Internal_{QO} = su(2) = so(3). \tag{8.27}$$

That's 66+64+3 = 133 elements altogether, and we make the identification:

$$MS_{QO} = LE_7. \tag{8.28}$$

O ⊗ O

Let e_{La} and e'_{Lb} be distinct and mutually commuting bases for the hypercomplex octonions. $External_{OO}$ is spanned by ω, $e_{La}\alpha$, $e'_{La}\beta$ (1-vector basis for nonuniversal $\mathbf{R}_{0,15}$), and $e_{La}\beta$, $e'_{La}\alpha$, $e_{La}e'_{Lb}\omega$, $e_{Lab}\epsilon$, $e'_{Lab}\epsilon$ (2-vectors).

$$External_{OO} = so(16). \tag{8.29}$$

$Spinor_{OO}$ is 128-dimensional, and again because $\mathbf{O}_L = \mathbf{O}_R$, $Internal_{RO}$ is empty. That's 120+128 = 248 elements altogether, and we make the identification:

$$MS_{OO} = LE_8. \tag{8.30}$$

Part of the significance of these constructions is the manner in which they parallel the construction of a physical model in the earlier chapters. In particular, that model was based on $\mathbf{T}_L(2) \otimes \mathbf{P}_R$ acting on $\mathbf{T}^2$, with $\mathbf{T}_L(2)$ identified as 'external', and $\mathbf{P}_R$ 'internal'. Some important distinctions can be made, however. $\mathbf{T}_L(2)$ can be viewed as either the complexification of $\mathbf{R}_{1,9}$, or as the Clifford algebra $\mathbf{R}_{1,10}$. For example, the elements

$$i\omega,\ q_{Lj}\alpha,\ e_{La}\beta, \tag{8.31}$$

are a 1-vector basis for $\mathbf{R}_{1,10} \simeq \mathbf{T}_L(2)$ (see the $\mathbf{Q} \otimes \mathbf{O}$ case above; this is only a slight modification of the 1-vector basis for $\mathbf{R}_{0,11}$ used there). For that matter, the set

$$i\omega,\ q_{Lj}\alpha,\ e_{Lp}\beta,\ p = 1, ..., 6, \tag{8.32}$$

is another 1-vector basis for $\mathbf{R}_{1,9}$. Why is the 1-vector basis given in 4.4 preferable to this? Perhaps it isn't, but there is a difference. The former basis is built up from a "Pauli" algebra underlying the (1,9)-Dirac algebra (see 6.1), just as the (1,3)-Dirac algebra is built up from the conventional Pauli algebra. The basis 8.31 is not so constructed. It should be possible to generate a basis for the Lie algebra of E_7 starting from the 1-vector basis for $\mathbf{R}_{0,11}$,

$$i\alpha,\ i\beta,\ e_{L7}q_{Lj}\omega,\ ie_{Lp}\omega,\ p = 1, ..., 6 \tag{8.33}$$

(a slight modification of the 1-vector basis for $\mathbf{R}_{1,9}$ in 4.4, where the basis above is founded on the "Pauli" algebra 6.1). To this add the basis of corresponding 2-vectors, completing the external subalgebra, $so(12)$. The internal space should be the same (ie., we are ignoring the presence of the complex unit i in 8.33), $su(2)$. The spinor space should be half (a Majorana doublet) of $\mathbf{T}^2$, where the other half is obtained by multiplication by i, and each half is closed under the action of $so(12) \times su(2)$). This construction of LE_7, however, is not valid for all of the other algebras of the magic square.

Questions arise from these ideas, for which there are presently no answers. Do these constructions signify a connection of the **T**-model to some GUT (for example, the E_6 GUT)? Or is the attractiveness of GUTs accidental, arising from algebraic similarities to the **T**-model? In either case, how may we naturally determine the global structure of the various manifold bits of the **T**-model?

8.3. $Spinor_{KK'}$ Outer Products.

In the next section we shall define commutator products for the MS_{RO} case (LF_4). As a first step in that direction we define the outer product of spinors. Three cases are developed here: **C**, **Q**, and **O**. Only the last is relevant. This is because in this section I am treating 2-dimensional **C** and 4-dimensional **Q** as spinor spaces for $\mathbf{R}(2)$ and $\mathbf{R}(4)$, but $\mathbf{C}_L \neq \mathbf{R}(2)$, and $\mathbf{Q}_L \neq \mathbf{R}(4)$ (the **Q** spinor outer product below is valid for $\mathbf{Q}_A \simeq \mathbf{R}(4)$ (see 2.28), in which case 4-dimensional **Q** is the complete spinor space). On the other hand, 8-dimensional **O** is the full spinor space for $\mathbf{O}_L$, because $\mathbf{O}_L \simeq \mathbf{R}(8)$. The first two cases establish the method.

C OUTER PRODUCTS

Consider the example of $\mathbf{C}_L$ acting on the spinor space **C**. **C** is 2-dimensional over **R** and as a real column spinor space its elements can be represented by

$$1 \rightarrow \begin{bmatrix} 1 \\ 0 \end{bmatrix}, \quad i \rightarrow \begin{bmatrix} 0 \\ 1 \end{bmatrix}. \tag{8.34}$$

This leads naturally to the representations

$$1_L \rightarrow \begin{bmatrix} 1 & 0 \\ 0 & 1 \end{bmatrix}, \quad i_L \rightarrow \begin{bmatrix} 0 & -1 \\ 1 & 0 \end{bmatrix}. \tag{8.35}$$

We expect the outer product of 1 with its transpose to be representable as

$$|1><1| \rightarrow \begin{bmatrix} 1 \\ 0 \end{bmatrix} \begin{bmatrix} 1 & 0 \end{bmatrix} = \begin{bmatrix} 1 & 0 \\ 0 & 0 \end{bmatrix}, \tag{8.36}$$

which is not a linear combination of the matrices 8.35. We need to expand the set by

$$1_L^* \rightarrow \begin{bmatrix} 1 & 0 \\ 0 & -1 \end{bmatrix}, \quad i_L^* \rightarrow \begin{bmatrix} 0 & 1 \\ 1 & 0 \end{bmatrix}, \tag{8.37}$$

the former just complex conjugation, the latter complex conjugation followed by i_L (see 2.30). Therefore,

$$R_C = |1><1| = \frac{1}{2}(1_L + 1_L^*), \tag{8.38}$$

which is the projector of the real part of **C**. That is

$$\frac{1}{2}(1_L + 1_L^*)[x] = Re(x) \tag{8.39}$$

for all $x \in \mathbf{C}$.

Any element x of **C** can be mapped to an element x_L of $\mathbf{C}_L$ by appending the subscript 'L'. We may now define a general outer product of x and y in **C**:

$$|x><y| = x_L R_C y_L^\dagger, \tag{8.40}$$

where $y^\dagger = y^*$ in this case. For example,

$$|1><i| \rightarrow \begin{bmatrix} 1 \\ 0 \end{bmatrix} \begin{bmatrix} 0 & 1 \end{bmatrix} = \begin{bmatrix} 0 & 1 \\ 0 & 0 \end{bmatrix}, \tag{8.41}$$

so we expect $|1><i|$ to map i $(= |i>)$ to 1 $(= |1>)$, and to map 1 to zero. Using 8.41,

$$|1><i| = 1_L R_C(-i_L) = -\frac{1}{2}(1_L + 1_L^*)i_L = \frac{1}{2}(i_L^* - i_L), \tag{8.42}$$

and

$$\frac{1}{2}(i_L^* - i_L)[i] = \frac{1}{2}(i(-i) - i(i)) = 1,$$
$$\frac{1}{2}(i_L^* - i_L)[1] = \frac{1}{2}(i(1) - i(1)) = 0,$$

as expected.

Equation 8.42 has exact counterparts when we replace **C** by **Q** or **O**. The problem is to define R_Q and R_O, which project the real parts of those algebras ($R_O = P_0$; see sections 2.4 and 7.1).

Q OUTER PRODUCTS

In the case of $\mathbf{Q}_L$ and $\mathbf{Q}_R$ acting on the spinor space **Q**, if we represent the basis for **Q** by

$$1 \to \begin{bmatrix} 1 \\ 0 \\ 0 \\ 0 \end{bmatrix}, \quad q_1 \to \begin{bmatrix} 0 \\ 1 \\ 0 \\ 0 \end{bmatrix}, \quad q_2 \to \begin{bmatrix} 0 \\ 0 \\ 1 \\ 0 \end{bmatrix}, \quad q_3 \to \begin{bmatrix} 0 \\ 0 \\ 0 \\ 1 \end{bmatrix}, \tag{8.43}$$

then

$$\begin{gathered}
q_{L1} \to \begin{bmatrix} 0 & -1 & 0 & 0 \\ 1 & 0 & 0 & 0 \\ 0 & 0 & 0 & -1 \\ 0 & 0 & 1 & 0 \end{bmatrix}, q_{L2} \to \begin{bmatrix} 0 & 0 & -1 & 0 \\ 0 & 0 & 0 & 1 \\ 1 & 0 & 0 & 0 \\ 0 & -1 & 0 & 0 \end{bmatrix}, \\
q_{L3} \to \begin{bmatrix} 0 & 0 & 0 & -1 \\ 0 & 0 & -1 & 0 \\ 0 & 1 & 0 & 0 \\ 1 & 0 & 0 & 0 \end{bmatrix}, \\
q_{R1} \to \begin{bmatrix} 0 & -1 & 0 & 0 \\ 1 & 0 & 0 & 0 \\ 0 & 0 & 0 & 1 \\ 0 & 0 & -1 & 0 \end{bmatrix}, q_{R2} \to \begin{bmatrix} 0 & 0 & -1 & 0 \\ 0 & 0 & 0 & -1 \\ 1 & 0 & 0 & 0 \\ 0 & 1 & 0 & 0 \end{bmatrix}, \\
q_{R3} \to \begin{bmatrix} 0 & 0 & 0 & -1 \\ 0 & 0 & 1 & 0 \\ 0 & -1 & 0 & 0 \\ 1 & 0 & 0 & 0 \end{bmatrix},
\end{gathered} \tag{8.44}$$

so that

$$-q_{L1}q_{R1} \rightarrow \begin{bmatrix} 1 & 0 & 0 & 0 \\ 0 & 1 & 0 & 0 \\ 0 & 0 & -1 & 0 \\ 0 & 0 & 0 & -1 \end{bmatrix}, -q_{L2}q_{R2} \rightarrow \begin{bmatrix} 1 & 0 & 0 & 0 \\ 0 & -1 & 0 & 0 \\ 0 & 0 & 1 & 0 \\ 0 & 0 & 0 & -1 \end{bmatrix},$$

$$-q_{L3}q_{R3} \rightarrow \begin{bmatrix} 1 & 0 & 0 & 0 \\ 0 & -1 & 0 & 0 \\ 0 & 0 & -1 & 0 \\ 0 & 0 & 0 & 1 \end{bmatrix}. \tag{8.45}$$

Therefore,

$$R_Q = \frac{1}{4}(1_L - \sum_{j=1}^{3} q_{Lj}q_{Rj}) \rightarrow \begin{bmatrix} 1 & 0 & 0 & 0 \\ 0 & 0 & 0 & 0 \\ 0 & 0 & 0 & 0 \\ 0 & 0 & 0 & 0 \end{bmatrix}, \tag{8.46}$$

and for all x and y in $\mathbf{Q}$ we define

$$|x >< y| = x_L R_Q y_L^{\dagger}. \tag{8.47}$$

As an example,

$$|q_1 >< q_2| = q_{L1}R_Q(-q_{L2}) \rightarrow \begin{bmatrix} 0 & 0 & 0 & 0 \\ 0 & 0 & 1 & 0 \\ 0 & 0 & 0 & 0 \\ 0 & 0 & 0 & 0 \end{bmatrix}.$$

This should map $q_2 \rightarrow q_1$, and everything else to zero.

$$-q_{L1}R_Q q_{L2}[1] = -q_{L1}R_Q[q_2] = -q_{L1}[Re(q_2)] = 0,$$

$$-q_{L1}R_Q q_{L2}[q_1] = -q_{L1}R_Q[-q_3] = -q_{L1}[Re(-q_3)] = 0,$$

$$-q_{L1}R_Q q_{L2}[q_2] = -q_{L1}R_Q[-1] = -q_{L1}[Re(-1)] = -q_{L1}[-1] = q_1,$$

$$-q_{L1}R_Q q_{L2}[q_3] = -q_{L1}R_Q[q_1] = -q_{L1}[Re(q_1)] = 0.$$

O OUTER PRODUCTS

In the case of the octonions,

$$R_O = P_0 = \frac{1}{8}(1_L + e_{L476} + e_{L517} + e_{L621} + e_{L732} + e_{L143} + e_{L254} + e_{L365}) \quad (8.48)$$

(see sections 2.4 and 7.1), and for all $x, y \in \mathbf{O}$,

$$|x >< y| = x_L P_0 y_L^\dagger. \quad (8.49)$$

Therefore, for example,

$$|e_a >< e_a| = -e_{La} P_0 e_{La} = P_a. \quad (8.50)$$

And the vector $so(8)$ generators in 7.102 can be written,

$$|e_b >< e_a| - |e_a >< e_b| = e_{Lb} P_0 e_{La}^\dagger - e_{La} P_0 e_{Lb}^\dagger. \quad (8.51)$$

For $a, b \in \{0, ..., 7\}$, this yields all 28 of the vector $so(8)$ generators on the last line of 7.102.

Note that if $X \in \mathbf{T}$, then

$$\begin{aligned} &R_O[X] \in \mathbf{C} \otimes \mathbf{Q}, \\ &R_Q R_O[X] \in \mathbf{C}, \\ &R_C R_Q R_O[X] \in \mathbf{R}. \end{aligned} \quad (8.52)$$

So the inner product

$$< X, Y >= R_C R_Q R_O[X^\dagger Y],$$

which is $SO(64)$ invariant. This expression for the inner product, however, hides the $U(1) \times SU(2) \times SU(3)$ symmetry that is manifested by constructing the inner product using projection operators from $\mathbf{T}$, the spinor space, instead of $\mathbf{T}_A$.

8.4. $LF_4 \simeq MS_{RO}$.

A basis for $External_{RO} \simeq so(9)$ consists of the 36 elements

$$\omega,\ e_{La}\beta,\ e_{La}\alpha,\ e_{Lab}\epsilon. \tag{8.53}$$

Denote these elements J_f, $f = 1, ..., 36$ (the order is immaterial). $Spinor_{RO}$ is just $\mathbf{O}^2$, 2×1 matrices over $\mathbf{O}$. Together $External_{RO}$ and $Spinor_{RO}$ define a representation of LF_4 ($Internal_{RO}$ is empty). Let $A, B \in External_{RO}$, and $X, Y \in Spinor_{RO}$. The commutator

$$[A, B] = AB - BA, \tag{8.54}$$

as expected. *External/Spinor* commutators are also simply defined:

$$[A, X] = -[X, A] = A[X]. \tag{8.55}$$

It only remains to define *Spinor/Spinor* commutators. Following [Baaklini], we define

$$[X, Y] = \frac{1}{4}\sum_{f=1}^{36} < X, J_f[Y] > J_f \tag{8.56}$$

(note: the spinor conjugation used in the definition of the inner product is in this case ordinary Hermitian transposition).

We can gain some insight into 8.57 by considering a few examples. Start with

$$[\begin{bmatrix} e_1 \\ 0 \end{bmatrix}, \begin{bmatrix} 1 \\ 0 \end{bmatrix}] = \frac{1}{4}\sum_{f=1}^{36} < \begin{bmatrix} e_1 \\ 0 \end{bmatrix}, J_f \begin{bmatrix} 1 \\ 0 \end{bmatrix} > J_f. \tag{8.57}$$

For the inner product on the righthand side of 8.58 to be nonzero, we must have

$$J_f \begin{bmatrix} 1 \\ 0 \end{bmatrix} = \begin{bmatrix} \pm e_1 \\ 0 \end{bmatrix}.$$

This occurs for $J_f = e_{L1}\alpha,\ e_{L26}\epsilon,\ e_{L34}\epsilon,\ e_{L57}\epsilon$. Therefore

$$\begin{aligned} [\begin{bmatrix} e_1 \\ 0 \end{bmatrix}, \begin{bmatrix} 1 \\ 0 \end{bmatrix}] &= \tfrac{1}{4}(e_{L1}\alpha + e_{L26}\epsilon + e_{L34}\epsilon + e_{L57}\epsilon) \\ &= \tfrac{1}{4}(e_{L1} + e_{L26} + e_{L34} + e_{L57})\alpha^{\oplus} \\ &\quad +\tfrac{1}{4}(-e_{L1} + e_{L26} + e_{L34} + e_{L57})\alpha^{\ominus}, \end{aligned} \tag{8.58}$$

where

$$\alpha^{\oplus} = \frac{1}{2}(\epsilon + \alpha) = \begin{bmatrix} 1 & 0 \\ 0 & 0 \end{bmatrix},$$

$$\alpha^{\ominus} = \frac{1}{2}(\epsilon - \alpha) = \begin{bmatrix} 0 & 0 \\ 0 & 1 \end{bmatrix}.$$

The coefficient of $\alpha^{\oplus}$, using table i.1 in appendix i, is just the outer product

$$\begin{aligned} \frac{1}{4}(e_{L1} + e_{L26} + e_{L34} + e_{L57}) &= |e_1 >< 1| - |1 >< e_1| \\ &= e_{L1}P_0 - P_0 e_{L1}^{\dagger} \\ &= e_{L1}P_0 + P_0 e_{L1}. \end{aligned} \tag{8.59}$$

The coefficient of $\alpha^{\ominus}$ is recognized to be (see 2.20)

$$\frac{1}{4}(-e_{L1} + e_{L26} + e_{L34} + e_{L57}) = \frac{1}{2}e_{R1}. \tag{8.60}$$

Likewise, it is simply determined using table i.6 in appendix i, that

$$\begin{aligned} [\begin{bmatrix} e_1 \\ 0 \end{bmatrix}, \begin{bmatrix} e_2 \\ 0 \end{bmatrix}] &= \frac{1}{4}(-e_{L6}\alpha + e_{L74}\epsilon - e_{L12}\epsilon + e_{L35}\epsilon) \\ &= \frac{1}{4}(-e_{L6} + e_{L74} - e_{L12} + e_{L35})\alpha^{\oplus} \\ &\quad + \frac{1}{4}(e_{L6} + e_{L74} - e_{L12} + e_{L35})\alpha^{\ominus}. \end{aligned} \tag{8.61}$$

In this case the coefficient of $\alpha^{\oplus}$ is

$$\begin{aligned} \frac{1}{4}(-e_{L6} + e_{L74} - e_{L12} + e_{L35}) &= |e_1 >< e_2| - |e_2 >< e_1| \\ &= e_{L1}P_0 e_{L2}^{\dagger} - e_{L2}P_0 e_{L1}^{\dagger} \\ &= e_{R1}P_0 e_{R2}^{\dagger} - e_{R2}P_0 e_{R1}^{\dagger}, \end{aligned} \tag{8.62}$$

and the coefficient of $\alpha^{\ominus}$ is

$$\frac{1}{4}(e_{L6} + e_{L74} - e_{L12} + e_{L35}) = \frac{1}{2}e_{R12}, \tag{8.63}$$

where

$$e_{R12}[x] = (xe_1)e_2 = e_{R2}e_{R1}[x].$$

The pattern is clear. In general,

$$\begin{aligned}[\begin{bmatrix} x \\ 0 \end{bmatrix}, \begin{bmatrix} y \\ 0 \end{bmatrix}] &= (x_R P_0 y_R^\dagger - y_R P_0 x_R^\dagger)\alpha^\oplus \\ &+ \frac{1}{4}(x_R y_R^\dagger - y_R x_R^\dagger)\alpha^\ominus, \end{aligned} \tag{8.64}$$

where if $x = \sum_0^7 x^a e_a \in \mathbf{O}$, then $x_L = \sum_0^7 x^a e_{La}$, and $x_R = \sum_0^7 x^a e_{Ra}$.

Note: From the results of section 7.3 we see that the coefficient of $\alpha^\oplus$ is vector $so(8)$, and of $\alpha^\ominus$ is spinor $so(8)$. That this should be so is related to the fact that F_4 is the automorphism group of the Jordan algebra of 3×3 Hermitian matrices over $\mathbf{O}$, which is also intimately linked to the notion of triality. Indeed, note that by 7.113 and 7.119, under the triality map Tri, if $a, b \in \{1, ..., 7\}$,

$$\begin{aligned} Tri\{\tfrac{1}{2} e_{R0} e_{Ra}^\dagger\} &= Tri\{-\tfrac{1}{2} e_{Ra}\} \\ &= e_{La} P_0 + P_0 e_{La} \\ &= e_{La} P_0 e_{L0}^\dagger - e_{L0} P_0 e_{La}^\dagger, \end{aligned}$$

and

$$\begin{aligned} Tri\{\tfrac{1}{2} e_{Ra} e_{Rb}^\dagger\} &= Tri\{-\tfrac{1}{2} e_{Ra} e_{Rb}\} \\ &= Tri\{\tfrac{1}{2} e_{Rab}\} \\ &= -e_{La} P_0 e_{Lb} + e_{Lb} P_0 e_{La} \\ &= e_{La} P_0 e_{Lb}^\dagger - e_{Lb} P_0 e_{La}^\dagger. \end{aligned}$$

Therefore, in general,

$$Tri\{\frac{1}{4} x_R y_R^\dagger\} = -Tri\{\frac{1}{4} y_R x_R^\dagger\} = \frac{1}{2}(y_R P_0 x_R^\dagger - x_R P_0 y_R^\dagger). \tag{8.65}$$

So

$$Tri\{\frac{1}{4}(x_R y_R^\dagger - y_R x_R^\dagger)\} = -(x_R P_0 y_R^\dagger - y_R P_0 x_R^\dagger). \tag{8.66}$$

That is, the coefficient of $\alpha^\oplus$ in 8.64 is minus the triality rotation of the coefficient of $\alpha^\ominus$.

In like manner we determine

$$[\begin{bmatrix} 0 \\ x \end{bmatrix}, \begin{bmatrix} 0 \\ y \end{bmatrix}] = (x_R P_0 y_R^\dagger - y_R P_0 x_R^\dagger)\alpha^\ominus + \frac{1}{4}(x_R y_R^\dagger - y_R x_R^\dagger)\alpha^\oplus. \tag{8.67}$$

Now consider the commutator

$$[\begin{bmatrix} e_1 \\ 0 \end{bmatrix}, \begin{bmatrix} 0 \\ 1 \end{bmatrix}] = \frac{1}{4}\sum_{f=1}^{36} < \begin{bmatrix} e_1 \\ 0 \end{bmatrix}, J_f \begin{bmatrix} 0 \\ 1 \end{bmatrix} > J_f. \tag{8.68}$$

In this case the inner product on the righthand side of 8.68 is nonzero only if

$$J_f \begin{bmatrix} 0 \\ 1 \end{bmatrix} = \begin{bmatrix} \pm e_1 \\ 0 \end{bmatrix}.$$

This occurs in the single case $J_f = e_{L1}\beta$. Hence,

$$[\begin{bmatrix} e_1 \\ 0 \end{bmatrix}, \begin{bmatrix} 0 \\ 1 \end{bmatrix}] = \frac{1}{4} e_{L1}\beta. \tag{8.69}$$

Similarly,

$$< \begin{bmatrix} e_2 \\ 0 \end{bmatrix}, J_f \begin{bmatrix} 0 \\ e_6 \end{bmatrix} > \neq 0$$

only if $J_f = e_{L1}\beta$. Since $e_{L1}[e_6] = -e_2$,

$$[\begin{bmatrix} e_2 \\ 0 \end{bmatrix}, \begin{bmatrix} 0 \\ e_6 \end{bmatrix}] = -\frac{1}{4} e_{L1}\beta. \tag{8.70}$$

Therefore, using the identities of section 7.1, if $a, b \in \{0, ..., 7\}$ are distinct, then

$$\begin{aligned} [\begin{bmatrix} e_a \\ 0 \end{bmatrix}, \begin{bmatrix} 0 \\ e_c \end{bmatrix}] &= \tfrac{1}{4}[e_{Ra}(2P_0 - \tfrac{1}{2})e_{Rb}^\dagger - e_{Rb}(2P_0 - \tfrac{1}{2})e_{Ra}^\dagger]\beta \\ &= [\tfrac{1}{2}e_{Ra}P_0 e_{Rb}^\dagger - \tfrac{1}{2}e_{Rb}P_0 e_{Ra}^\dagger - \tfrac{1}{8}(e_{Ra}e_{Rb}^\dagger - e_{Rb}e_{Ra}^\dagger)]\beta. \end{aligned} \tag{8.71}$$

Finally, for all $a \in \{0, ..., 7\}$,

$$[\begin{bmatrix} e_a \\ 0 \end{bmatrix}, \begin{bmatrix} 0 \\ e_a \end{bmatrix}] = \frac{1}{4}\omega. \tag{8.72}$$

Define

$$\beta^{\oplus} = \begin{bmatrix} 0 & 1 \\ 0 & 0 \end{bmatrix}, \ \beta^{\ominus} = \begin{bmatrix} 0 & 0 \\ 1 & 0 \end{bmatrix}. \tag{8.73}$$

Combining these results, we see in general that

$$\begin{aligned}
[\begin{bmatrix} x \\ 0 \end{bmatrix}, \begin{bmatrix} 0 \\ y \end{bmatrix}] &= \tfrac{1}{4}[x_R(2P_0 - \tfrac{1}{2})y_R^{\dagger} - y_R(2P_0 - \tfrac{1}{2})x_R^{\dagger}]\beta + \tfrac{1}{8}[x_R y_R^{\dagger} + y_R x_R^{\dagger}]\omega \\
&= \{\tfrac{1}{4}[x_R(2P_0 - \tfrac{1}{2})y_R^{\dagger} - y_R(2P_0 - \tfrac{1}{2})x_R^{\dagger}] + \tfrac{1}{8}[x_R y_R^{\dagger} + y_R x_R^{\dagger}]\}\beta^{\oplus} \\
&\quad + \{\tfrac{1}{4}[x_R(2P_0 - \tfrac{1}{2})y_R^{\dagger} - y_R(2P_0 - \tfrac{1}{2})x_R^{\dagger}] - \tfrac{1}{8}[x_R y_R^{\dagger} + y_R x_R^{\dagger}]\} \\
&= \{\tfrac{1}{2}(x_R P_0 y_R^{\dagger} - y_R P_0 x_R^{\dagger}) + \tfrac{1}{4} y_R x_R^{\dagger}\}\beta^{\oplus} \\
&\quad + \{\tfrac{1}{2}(x_R P_0 y_R^{\dagger} - y_R P_0 x_R^{\dagger}) - \tfrac{1}{4} x_R y_R^{\dagger}\}\beta^{\ominus} \\
&= \tfrac{1}{4}(xy^{\dagger})_L\beta^{\oplus} - \tfrac{1}{4}(yx^{\dagger})_L\beta^{\ominus}.
\end{aligned} \tag{8.74}$$

Fiddling with all these commutators leads to a unified expression of a general LF_4 spinor commutator:

$$\boxed{\begin{aligned}
[\begin{bmatrix} x \\ y \end{bmatrix}, \begin{bmatrix} u \\ v \end{bmatrix}] &= \{\tfrac{1}{2}\begin{bmatrix} x_R P_0 \\ y_R P_0 \end{bmatrix}[\ P_0 u_R^{\dagger} \quad P_0 v_R^{\dagger}\ ] - (same)^{\dagger}\} + \{same\}^T \\
&\quad + \tfrac{1}{4}\begin{bmatrix} -y_R \\ x_R \end{bmatrix}[\ -v_R^{\dagger} \quad u_R^{\dagger}\ ] - (same)^{\dagger},
\end{aligned}} \tag{8.75}$$

where A^T is the transpose of the matrix A. Taking advantage of the triality map, we may rewrite 8.75,

$$\boxed{\begin{aligned}
[\begin{bmatrix} x \\ y \end{bmatrix}, \begin{bmatrix} u \\ v \end{bmatrix}] &= \tfrac{1}{4}\begin{bmatrix} -y_R \\ x_R \end{bmatrix}[\ -v_R^{\dagger} \quad u_R^{\dagger}\ ] - \tfrac{1}{4}\begin{bmatrix} -v_R \\ u_R \end{bmatrix}[\ -y_R^{\dagger} \quad x_R^{\dagger}\ ] \\
&\quad - Tri\{\tfrac{1}{4}\begin{bmatrix} x_R \\ y_R \end{bmatrix}[\ u_R^{\dagger} \quad v_R^{\dagger}\ ] - \tfrac{1}{4}\begin{bmatrix} u_R \\ v_R \end{bmatrix}[\ x_R^{\dagger} \quad y_R^{\dagger}\ ]\}.
\end{aligned}} \tag{8.76}$$

8.5. $J_3^{\mathbf{O}}$ and F_4.

Let $J_3^{\mathbf{O}}$ be the set of all 3×3 Hermitian matrices over the octonions (a 27-dimensional real vector space). This set is closed under the symmetric Jordan product

$$X \circ Y = \frac{1}{2}(XY + YX), \tag{8.77}$$

for all $X, Y \in J_3^{\mathbf{O}}$, and in addition,

$$X \circ (Y \circ X^2) = (X \circ Y) \circ X^2. \tag{8.78}$$

These conditions make $J_3^{\mathbf{O}}$ a Jordan algebra [9], a class of algebras initially studied for their connection to quantum theory. $J_3^{\mathbf{O}}$ is *the* exceptional Jordan algebra, the only one not representable (given the product 8.77) by a set of associative matrices (another in a growing list of exceptional mathematical objects associated with $\mathbf{O}$).

The automorphism group of $J_3^{\mathbf{O}}$ is F_4, and in this section we will construct a representation of its Lie algebra, LF_4, that will look quite different from that developed in the previous section.

Let

$$X = \begin{bmatrix} a & z & y^\dagger \\ z^\dagger & b & x \\ y & x^\dagger & c \end{bmatrix} \in J_3^{\mathbf{O}}, \tag{8.79}$$

$a, b, c \in \mathbf{R}$, and $x, y, z \in \mathbf{O}$. Clearly $X^\dagger = X$. Let Y be another element of $J_3^{\mathbf{O}}$. Since

$$\begin{aligned} X \circ Y &= \tfrac{1}{2}(XY + YX) \\ &= \tfrac{1}{4}((X+Y)^2 - X^2 - Y^2), \end{aligned}$$

F_4 consists of those maps M on $J_3^{\mathbf{O}}$ satisfying

$$M[X^2] = (M[X])^2 \tag{8.80}$$

for all $X \in J_3^{\mathbf{O}}$. Therefore, LF_4 consists of those maps m on $J_3^{\mathbf{O}}$ satisfying

$$\boxed{m[X^2] = (m[X])X + X(m[X]).} \tag{8.81}$$

The most obvious of such maps arise from the $so(3)$ subalgebra of LF_4 arising from commutation of X in 8.79 with the $so(3)$ basis

$$\begin{bmatrix} 0 & 1 & 0 \\ -1 & 0 & 0 \\ 0 & 0 & 0 \end{bmatrix}, \begin{bmatrix} 0 & 0 & -1 \\ 0 & 0 & 0 \\ 1 & 0 & 0 \end{bmatrix}, \begin{bmatrix} 0 & 0 & 0 \\ 0 & 0 & 1 \\ 0 & -1 & 0 \end{bmatrix}. \tag{8.82}$$

If u is a 3×3 real matrix linear in the matrices 8.82, and $U = exp(u)$, then the $SO(3)$ action of the form

$$X \to UXU^{\dagger} = UXU^{-1} \tag{8.83}$$

clearly maps $J_3^{\mathbf{O}} \to J_3^{\mathbf{O}}$ and is an $J_3^{\mathbf{O}}$ automorphism. The matrices 8.82 account for 3 of the 52 dimensions of LF_4.

The remaining 49 arise from octonion actions and can be most easily constructed using the triality map Tri defined in section 7.3. Recall that the set

$$\{e_{La}, e_{Lbc} : a, b, c \in \{1, ..., 7\}\} \tag{8.84}$$

forms a basis for a representation of $so(8)$, and that if g_0 is linear in the elements 8.84, then

$$g_0,\ g_1 = Tri\{g_0\},\ g_2 = Tri^2\{g_0\} = Tri\{g_1\} \tag{8.85}$$

form a triality triple of $so(8)$ actions on $\mathbf{O}$ satisfying Freudenthal's principle of triality,

$$T_0 g_i T_0[xy] = (g_{i+1}[x])y + x(g_{i+2}[y]), \tag{8.86}$$

the indices taken modulo 3, from 0 to 2.

Now define the following action on $X \in J_3^{\mathbf{O}}$:

$$\begin{aligned} m[X] &= \begin{bmatrix} 0 & g_2[z] & (g_1[y])^{\dagger} \\ ... & 0 & g_0[x] \\ ... & ... & 0 \end{bmatrix} \\ &= \begin{bmatrix} 0 & g_2[z] & T_0 g_1 T_0[y^{\dagger}] \\ ... & 0 & g_0[x] \\ ... & ... & 0 \end{bmatrix}, \end{aligned} \tag{8.87}$$

where for the rest of this section, ... in a matrix will indicate that what belongs in that position is the Hermitian conjugate of the term across the principal diagonal. Then

$$
\begin{aligned}
&(m[X])X + X(m[X]) \\
&= \begin{bmatrix} (g_2[z])z^\dagger + (g_1[y])^\dagger y & g_2[z]b + (g_1[y])^\dagger x^\dagger & --- \\ \cdots & --- & --- \\ \cdots & \cdots & --- \end{bmatrix} \\
&+ \begin{bmatrix} z(g_2[z])^\dagger + y^\dagger(g_1[y]) & ag_2[z] + y^\dagger(g_0[x])^\dagger & --- \\ \cdots & --- & --- \\ \cdots & \cdots & --- \end{bmatrix}.
\end{aligned}
\tag{8.88}
$$

The one diagonal and one off-diagonal terms shown in 8.88 will prove sufficient to finish its construction. The rest of the terms are analysed similarly.

The diagonal term is

$$
\begin{aligned}
&(g_2[z])z^\dagger + z(g_2[z])^\dagger + (g_1[y])^\dagger y + y^\dagger(g_1[y]) \\
&= 2 < g_2[z]), z > +2 < g_1[y], y > \\
&= 0,
\end{aligned}
$$

since, for example, $g_2[z]$, an infinitesimal $so(8)$ rotation of z, is perpendicular to z. The off-diagonal term is

$$
\begin{aligned}
g_2[z](a+b) + (xg_1[y] + g_0[x]y)^\dagger &= g_2[z](a+b) + (T_0 g_2 T_0[xy])^\dagger \\
&= g_2[z](a+b) + T_0 T_0 g_2 T_0[xy] \\
&= g_2[z](a+b) + g_2[y^\dagger x^\dagger] \\
&= g_2[z(a+b) + y^\dagger x^\dagger].
\end{aligned}
\tag{8.89}
$$

Therefore,

$$
\begin{aligned}
&(m[X])X + X(m[X]) \\
&= \begin{bmatrix} 0 & g_2[z(a+b) + y^\dagger x^\dagger] & --- \\ \cdots & 0 & --- \\ \cdots & \cdots & 0 \end{bmatrix}.
\end{aligned}
\tag{8.90}
$$

The square of the matrix X is

$$X^2 = \begin{bmatrix} a^2 + zz^\dagger + y^\dagger y & z(a+b) + y^\dagger x^\dagger & --- \\ \cdots & --- & --- \\ \cdots & \cdots & --- \end{bmatrix}. \tag{8.91}$$

Therefore

$$\begin{aligned} m[X^2] &= \begin{bmatrix} 0 & g_2[z(a+b) + y^\dagger x^\dagger] & --- \\ \cdots & 0 & --- \\ \cdots & \cdots & 0 \end{bmatrix} \\ &= (m[X])X + X(m[X]). \end{aligned} \tag{8.92}$$

That is, the actions m defined in 8.87, the collection of which is 28-dimensional ($so(8)$), are elements of LF_4. Together with the the $so(3)$ actions defined above, this accounts for $3 + 28 = 31$ of the 52-dimensional LF_4.

The remaining 21 dimensions of LF_4 arise from $so(3)$ commutators with $so(8)$. Let S be the first of the matrices 8.82. Then the following commutator establishes the pattern for all $[so(3), so(8)]$ commutators:

$$\begin{aligned} & m[S,X] - [S, m[X]] \\ &= \begin{bmatrix} 0 & g_2[b-a] & T_0 g_1 T_0[x] \\ \cdots & 0 & -g_0[y^\dagger] \\ \cdots & \cdots & 0 \end{bmatrix} \\ & - \begin{bmatrix} g_2[z] + (g_2[z])^\dagger & 0 & g_0[x] \\ 0 & -(g_2[z] + (g_2[z])^\dagger) & -T_0 g_1 T_0[y^\dagger] \\ \cdots & \cdots & 0 \end{bmatrix} \\ &= \begin{bmatrix} -2P_0 g_2[z] & g_2[b-a] & (-g_0 + T_0 g_1 T_0)[x] \\ \cdots & 2P_0 g_2[z] & (-g_0 + T_0 g_1 T_0)[y^\dagger] \\ \cdots & \cdots & 0 \end{bmatrix}, \end{aligned} \tag{8.93}$$

since $x + x^\dagger = 2P_0[x]$.

Note that

$$\begin{aligned} T_0 Tri\{e_{La}\} T_0 &= T_0 e_{Ra} T_0 = -e_{La}, \\ T_0 Tri\{e_{Lab}\} T_0 &= -T_0 e_{Rab} T_0 = +e_{Lab}. \end{aligned} \tag{8.94}$$

So if we give g_0 the general form

$$g_0 = \frac{1}{2}\sum_1^7 g_0^a e_{La} + \frac{1}{2}\sum_1^7 g_0^{bc} e_{Lbc}, \tag{8.95}$$

g_0^a and g_0^{bc} real, then

$$-g_0 + T_0 g_1 T_0 = -\sum_1^7 g_0^a e_{La}. \tag{8.96}$$

Furthermore,

$$\begin{aligned} g_2 = Tri^2\{g_0\} &= \textstyle\sum_1^7 g_0^a(e_{La}P_0 + P_0 e_{La}) \\ &\quad + \textstyle\sum_1^7 g_0^{bc}(e_{Lb}P_0 e_{Lc} - e_{Lc}P_0 e_{Lb}). \end{aligned} \tag{8.97}$$

Therefore, since $P_0 e_{La} P_0 = 0$, $a = 1, ..., 7$, and $P_0^2 = P_0$,

$$P_0 g_2 = \sum_1^7 g_0^a P_0 e_{La}. \tag{8.98}$$

Also, if r is real, then $P_0 e_{La}[r] = 0$, and $P_0[r] = r$. This implies

$$g_2[b-a] = \sum_1^7 g_0^a e_{La}[b-a]. \tag{8.99}$$

Finally, using 8.96,98,99, we may rewrite 8.93

$$m[S,X] - [s, m[X]]$$

$$= \begin{bmatrix} -2\sum_1^7 g_0^a e_{La}[z] & \sum_1^7 g_0^a e_{La}[b-a] & -\sum_1^7 g_0^a e_{La}[x] \\ \cdots & 2\sum_1^7 g_0^a e_{La}[z] & -\sum_1^7 g_0^a e_{La}[y^\dagger] \\ \cdots & \cdots & 0 \end{bmatrix}. \tag{8.100}$$

That is, the commutator of the $so(3)$ action ($[S,X]$) with the $so(8)$ action ($m[X]$) is the 7-dimensional action 8.100. Carrying on in this way, commuting the other two $so(3)$ basis actions (associated with the other two generators in 8.82), yields similar results. Together with 8.100 that yields a 21-dimensional subspace of LF_4 generators. Since

$$3 + 28 + 21 = 52,$$

that completes this new basis for LF_4.

Notice that S, together with the 28 generators 8.87, and the subsequent 7 generators 8.100 resulting from commutation, form a 36-dimensional subalgebra of LF_4 isomorphic to $so(9) \simeq External_{RO}$.

8.6. More Magic Square.

The Lie algebra of the automorphism group of $J_3^{\mathbf{R}}$ (Jordan algebra of 3×3 real Hermitian matrices) is just $so(3)$, with basis given in 8.82; that of $J_3^{\mathbf{C}}$ (3×3 complex Hermitian matrices) is $su(3)$, with five additional generators

$$\begin{bmatrix} i & 0 & 0 \\ 0 & -i & 0 \\ 0 & 0 & 0 \end{bmatrix}, \begin{bmatrix} i & 0 & 0 \\ 0 & i & 0 \\ 0 & 0 & -2i \end{bmatrix},$$

$$\begin{bmatrix} 0 & i & 0 \\ i & 0 & 0 \\ 0 & 0 & 0 \end{bmatrix}, \begin{bmatrix} 0 & 0 & i \\ 0 & 0 & 0 \\ i & 0 & 0 \end{bmatrix}, \begin{bmatrix} 0 & 0 & 0 \\ 0 & 0 & i \\ 0 & i & 0 \end{bmatrix}. \tag{8.101}$$

The space of matrices $\mathbf{C}\otimes J_3^{\mathbf{O}}$ is the set of all 3×3 matrices over $\mathbf{C}\otimes\mathbf{O}$ satisfying

$$X^{\phi} = X, \tag{8.102}$$

where X^{ϕ} is transposition together with octonion conjugation. That is, if X^* is simple complex conjugation, no transposition, then

$$X^{\phi} = X^{*\dagger}, \tag{8.103}$$

where now

$$X = \begin{bmatrix} a & z & y^{\phi} \\ z^{\phi} & b & x \\ y & x^{\phi} & c \end{bmatrix}, \tag{8.104}$$

$a, b, c \in \mathbf{C}$, and $x, y, z \in \mathbf{C}\otimes\mathbf{O}$.

If $U \in SU(3)$ is generated from the matrices in 8.82,101, then the set $\mathbf{C}\otimes J_3^{\mathbf{O}}$ is invariant under the action

$$X \rightarrow UXU^{\phi} = UXU^{T}, \tag{8.105}$$

where U^T is the transpose of U ($U^T = U^{*\dagger}$). Of course, an action of the form 8.104 is not an automorphism of $\mathbf{C}\otimes J_3^{\mathbf{O}}$ given the product 8.77.

The infinitesimal (Lie algebra) version of the action 8.105 is commutation for the matrices 8.82, and anticommutation for the matrices in 8.101. Let R be the first of the matrices in 8.101, T the third (both nonzero in the upper left 2×2 block). Then the anticommutator

$$\{R, X\} = \begin{bmatrix} 2ia & 0 & iy^{\phi} \\ 0 & -2ib & -ix \\ iy & -ix^{\phi} & 0 \end{bmatrix}, \tag{8.106}$$

and

$$m\{R,X\} = \begin{bmatrix} 0 & 0 & T_O g_1 T_O[iy^\phi] \\ 0 & 0 & g_O[-ix] \\ \cdots & \cdots & 0 \end{bmatrix} = \{R, m[X]\} \tag{8.107}$$

(note that if $y \in \mathbf{C} \otimes \mathbf{O}$, then $T_O[y] = y^\phi$). That is,

$$m\{R,X\} - \{R, m[X]\} = 0; \tag{8.108}$$

the commutator of the action 8.106 arising from the diagonal $su(3)$ generator R, and the action 8.87, which is also in a sense diagonal (in that while x, y, z may be changed, they are not shifted from their matrix component slots), is zero.

On the other hand

$$m\{T,X\} - \{T, m[X]\}$$

$$= \begin{bmatrix} 0 & g_2[i(b+a)] & T_O g_1 T_O[ix] \\ \cdots & 0 & g_O[iy^\phi] \\ \cdots & \cdots & 0 \end{bmatrix}$$

$$- \begin{bmatrix} i(g_2[z] + (g_2[z])^\phi) & 0 & ig_O[x] \\ 0 & i(g_2[z] + (g_2[z])^\phi) & iT_O g_1 T_O[y^\phi] \\ \cdots & \cdots & 0 \end{bmatrix} \tag{8.109}$$

$$= \begin{bmatrix} 2iP_O g_2[z] & ig_2[b+a] & i(-g_O + T_O g_1 T_O)[x] \\ \cdots & 2iP_O g_2[z] & -i(-g_O + iT_O g_1 T_O)[y^\phi] \\ \cdots & \cdots & 0 \end{bmatrix}.$$

A quick comparison of this commutator with the commutator 8.93 should be enough to convince the reader that the commutator above, like 8.93, gives rise to 7 additional generators. The same will be true of the other two off-diagonal matrices in 8.101. And there the Lie algebra closes, with 28 elements from 8.87, 8 from 8.82,101, and then from the 6 off-diagonal elements of $su(3)$, $6 \times 7 = 42$ additional elements, for

$$28 + 8 + 42 = 78$$

all together. That is, once again from a combination of $\mathbf{C}$ and $\mathbf{O}$ we have arrived at LE_6.

Again note that the 28 elements 8.87, together with the generators R, S, T (the $su(2)$ subalgebra of $su(3)$ nonzero in the upper left 2×2 matrix block), and their commutators with $so(8)$ (8.87) (which give rise to 14 additional elements, 7 for the off-diagonal S, 7 for T), yields a subalgebra of LE_6 of order $28 + 3 + 14 = 45$ isomorphic to

$$so(10) \simeq External_{CO}.$$

LE_7 arises in the same way if we replace $\mathbf{C}$ by $\mathbf{Q}$. That is, we extend $so(3)$, the automorphism Lie algebra of $J_3^{\mathbf{R}}$, to $sp(3)$, the automorphism Lie algebra of $J_3^{\mathbf{Q}}$ (3×3 Hermitian matrices over $\mathbf{Q}$), by augmenting the 3 $so(3)$ generators 8.82 with the 18 additional generators

$$\begin{bmatrix} q_j & 0 & 0 \\ 0 & 0 & 0 \\ 0 & 0 & 0 \end{bmatrix}, \begin{bmatrix} 0 & 0 & 0 \\ 0 & q_j & 0 \\ 0 & 0 & 0 \end{bmatrix}, \begin{bmatrix} 0 & 0 & 0 \\ 0 & 0 & 0 \\ 0 & 0 & q_j \end{bmatrix},$$
$$\begin{bmatrix} 0 & q_j & 0 \\ q_j & 0 & 0 \\ 0 & 0 & 0 \end{bmatrix}, \begin{bmatrix} 0 & 0 & q_j \\ 0 & 0 & 0 \\ q_j & 0 & 0 \end{bmatrix}, \begin{bmatrix} 0 & 0 & 0 \\ 0 & 0 & q_j \\ 0 & q_j & 0 \end{bmatrix}. \tag{8.110}$$

If $U \in Sp(3)$ is generated from a linear combination of the matrices 8.82,110, then the action 8.105 on $X \in \mathbf{Q} \otimes J_3^{\mathbf{O}}$ maps this set to itself. Again, this translates to an anticommutator Lie algebra action on X arising from the extra generators 8.110.

Of these, 9 are diagonal, 9 off-diagonal, and as was the case with $su(3)$, commutators of the form 8.93,105 give rise to nonzero results only for the off-diagonal generators of $sp(3)$, and in each case the resultant space of new generators is 7-dimensional. This closes the algebra on the

$$28 + 21 + 84 = 133$$

elements of a new representation of LE_7.

Getting LE_8 from $\mathbf{O} \otimes J_3^{\mathbf{O}}$ is slightly trickier. In this case there are two distinct copies of $\mathbf{O}$ commuting with each other (denote them $\mathbf{O}_1$ and $\mathbf{O}_2$), and while $X \in \mathbf{O}_1 \otimes J_3^{\mathbf{O}_2}$ is still of the form 8.103, where $a, b, c \in \mathbf{O}_1$, and $x, y, z \in \mathbf{O}_1 \otimes \mathbf{O}_2$, the map $x \to x^{\phi}$ conjugates only the $\mathbf{O}_2$ part of x.

We begin again with the 28 $so(8)$ generators 8.87, and the 3 $so(3)$ generators 8.82. As before we expand the set of $so(3)$ generators. In the $\mathbf{C} \otimes J_3^{\mathbf{O}}$ case we expanded $so(3)$ to $su(3)$, the Lie algebra of the automorphism of $J_3^{\mathbf{C}}$; in the $\mathbf{Q} \otimes J_3^{\mathbf{O}}$ case we expanded $so(3)$ to $sp(3)$, the Lie algebra of the automorphism of $J_3^{\mathbf{Q}}$; in the $\mathbf{O}_1 \otimes J_3^{\mathbf{O}_2}$ case we expand $so(3)$ to LF_4, the Lie algebra of the automorphism of $J_3^{\mathbf{O}_1}$. That gives us 28 elements from $so(8)$, and 52 elements from LF_4 (which contains another distinct $so(8)$). Of the 52 generators of this new LF_4, 28 are diagonal in the sense that the maps 8.87 are diagonal, and 24 are off-diagonal. Commutators of the 28 diagonal generators (the $so(8)$ of $\mathbf{O}_1$) with the $so(8)$ of $\mathbf{O}_2$ yield nothing new, but each of the 24 off-diagonal generators gives rise to a 7-dimensional space of new generators. That yields,

$$28 + 52 + 168 = 248$$

generators all together, and the set closes here on LE_8. (The interested readers may carry out the details for themselves, as well as the identification of the $so(16) \simeq External_{OO}$ of LE_8 in this case.)

Appendix i. $\mathbf{O}_L$ Actions: Product Rule $e_a e_{a+1} = e_{a+5}$.

Given the octonion multiplication,

$$e_a e_{a+1} = e_{a+5},$$

the resulting multiplication table is

$$\begin{bmatrix} 1 & e_1 & e_2 & e_3 & e_4 & e_5 & e_6 & e_7 \\ e_1 & -1 & e_6 & e_4 & -e_3 & e_7 & -e_2 & -e_5 \\ e_2 & -e_6 & -1 & e_7 & e_5 & -e_4 & e_1 & -e_3 \\ e_3 & -e_4 & -e_7 & -1 & e_1 & e_6 & -e_5 & e_2 \\ e_4 & e_3 & -e_5 & -e_1 & -1 & e_2 & e_7 & -e_6 \\ e_5 & -e_7 & e_4 & -e_6 & -e_2 & -1 & e_3 & e_1 \\ e_6 & e_2 & -e_1 & e_5 & -e_7 & -e_3 & -1 & e_4 \\ e_7 & e_5 & e_3 & -e_2 & e_6 & -e_1 & -e_4 & -1 \end{bmatrix}. \qquad \text{(i.0)}$$

$\mathbf{O}$ is the object space of the 64-dimensional $\mathbf{O}_L$. In the tables listed below are the actions of the basis elements of $\mathbf{O}_L$ on the basis of $\mathbf{O}$.

As an example of how to use these tables, consider the bold-faced elements in table i.0. These indicate that

$$e_{L621}[e_3] = -e_3,$$

where the sign is drawn from the slot in the relevant row and column. As another example, from table i.1 we see that

$$e_{L34}[e_5] = -e_7,$$

where e_5 is on the left side, determining the row, e_7 is on the right at the end of the same row, and e_{L34} is on the top, determining the column. A minus sign is in the appropriate row and column, which is the sign of the action. The i.a table is associated with e_a, where $e_0 = 1$. Note that the adjoint elements along the top of table i.0 are the diagonal elements of $\mathbf{O}_L$, denoted I_a in 2.46. These were used to resolve the identity of $\mathbf{O}_L$ in section 2.4.

	1	e_{L476}	e_{L517}	$\mathbf{e_{L621}}$	e_{L732}	e_{L143}	e_{L254}	e_{L365}	
1	+	+	+	+	+	+	+	+	1
e_1	+	−	+	+	−	+	−	−	e_1
e_2	+	−	−	+	+	−	+	−	e_2
$\mathbf{e_3}$	+	−	−	−	+	+	−	+	$\mathbf{e_3}$
e_4	+	+	−	−	−	+	+	−	e_4
e_5	+	−	+	−	−	−	+	+	e_5
e_6	+	+	−	+	−	−	−	+	e_6
e_7	+	+	+	−	+	−	−	−	e_7

(i.1)

	e_{L1}	e_{L26}	e_{L34}	e_{L57}	e_{L546}	e_{L235}	e_{L427}	e_{L367}	
1	+	+	+	+	+	+	+	+	e_1
e_1	−	−	−	−	+	+	+	+	1
e_2	+	+	−	−	+	−	−	+	e_6
e_3	+	−	+	−	+	−	+	−	e_4
e_4	−	+	−	+	+	−	+	−	e_3
e_5	+	−	−	+	−	−	+	+	e_7
e_6	−	−	+	+	+	−	−	+	e_2
e_7	−	+	+	−	−	−	+	+	e_5

(i.2)

	e_{L2}	e_{L37}	e_{L45}	e_{L61}	e_{L657}	e_{L346}	e_{L531}	e_{L471}	
1	+	+	+	+	+	+	+	+	e_2
e_1	−	+	+	−	−	−	+	+	e_6
e_2	−	−	−	−	+	+	+	+	1
e_3	+	+	−	−	+	−	−	+	e_7
e_4	+	−	+	−	+	−	+	−	e_5
e_5	−	+	−	+	+	−	+	−	e_4
e_6	+	−	−	+	−	−	+	+	e_1
e_7	−	−	+	+	+	−	−	+	e_3

(i.3)

	e_{L3}	e_{L41}	e_{L56}	e_{L72}	e_{L761}	e_{L457}	e_{L642}	e_{L512}	
1	+	+	+	+	+	+	+	+	e_3
e_1	−	−	+	+	+	−	−	+	e_4
e_2	−	+	+	−	−	−	+	+	e_7
e_3	−	−	−	−	+	+	+	+	1
e_4	+	+	−	−	+	−	−	+	e_1
e_5	+	−	+	−	+	−	+	−	e_6
e_6	−	+	−	+	+	−	+	−	e_5
e_7	+	−	−	+	−	−	+	+	e_2

(i.4)

	e_{L4}	e_{L52}	e_{L67}	e_{L13}	e_{L172}	e_{L561}	e_{L753}	e_{L623}	
1	+	+	+	+	+	+	+	+	e_4
e_1	+	−	−	+	−	−	+	+	e_3
e_2	−	−	+	+	+	−	−	+	e_5
e_3	−	+	+	−	−	−	+	+	e_1
e_4	−	−	−	−	+	+	+	+	1
e_5	+	+	−	−	+	−	−	+	e_2
e_6	+	−	+	−	+	−	+	−	e_7
e_7	−	+	−	+	+	−	+	−	e_6

(i.5)

	e_{L5}	e_{L63}	e_{L71}	e_{L24}	e_{L213}	e_{L672}	e_{L164}	e_{L734}	
1	+	+	+	+	+	+	+	+	e_5
e_1	−	+	−	+	+	−	+	−	e_7
e_2	+	−	−	+	−	−	+	+	e_4
e_3	−	−	+	+	+	−	−	+	e_6
e_4	−	+	+	−	−	−	+	+	e_2
e_5	−	−	−	−	+	+	+	+	1
e_6	+	+	−	−	+	−	−	+	e_3
e_7	+	−	+	−	+	−	+	−	e_1

(i.6)

	e_{L6}	e_{L74}	e_{L12}	e_{L35}	e_{L324}	e_{L713}	e_{L275}	e_{L145}	
1	+	+	+	+	+	+	+	+	e_6
e_1	+	−	+	−	+	−	+	−	e_2
e_2	−	+	−	+	+	−	+	−	e_1
e_3	+	−	−	+	−	−	+	+	e_5
e_4	−	−	+	+	+	−	−	+	e_7
e_5	−	+	+	−	−	−	+	+	e_3
e_6	−	−	−	−	+	+	+	+	1
e_7	+	+	−	−	+	−	−	+	e_4

(i.7)

	e_{L7}	e_{L15}	e_{L23}	e_{L46}	e_{L435}	e_{L124}	e_{L316}	e_{L256}	
1	+	+	+	+	+	+	+	+	e_7
e_1	+	+	−	−	+	−	−	+	e_5
e_2	+	−	+	−	+	−	+	−	e_3
e_3	−	+	−	+	+	−	+	−	e_2
e_4	+	−	−	+	−	−	+	+	e_6
e_5	−	−	+	+	+	−	−	+	e_1
e_6	−	+	+	−	−	−	+	+	e_4
e_7	−	−	−	−	+	+	+	+	1

(i.8)

Appendix ii. $\mathbf{O}_R$ Actions: Product Rule $e_a e_{a+1} = e_{a+5}$.

For the sake of completeness, the actions of the basis of $\mathbf{O}_R$ on $\mathbf{O}$ are also given.

	1	e_{R476}	e_{R517}	e_{R621}	e_{R732}	e_{R143}	e_{R254}	e_{R365}	
1	+	+	+	+	+	+	+	+	1
e_1	+	−	+	+	−	+	−	−	e_1
e_2	+	−	−	+	+	−	+	−	e_2
e_3	+	−	−	−	+	+	−	+	e_3
e_4	+	+	−	−	−	+	+	−	e_4
e_5	+	−	+	−	−	−	+	+	e_5
e_6	+	+	−	+	−	−	−	+	e_6
e_7	+	+	+	−	+	−	−	−	e_7

(ii.0)

	e_{R1}	e_{R26}	e_{R34}	e_{R57}	e_{R564}	e_{R253}	e_{R472}	e_{R376}	
1	+	+	+	+	+	+	+	+	e_1
e_1	−	−	−	−	+	+	+	+	1
e_2	−	−	+	+	−	+	+	−	e_6
e_3	−	+	−	+	−	+	−	+	e_4
e_4	+	−	+	−	−	+	−	+	e_3
e_5	−	+	+	−	+	+	−	−	e_7
e_6	+	+	−	−	−	+	+	−	e_2
e_7	+	−	−	+	+	+	−	−	e_5

(ii.1)

	e_{R2}	e_{R37}	e_{R45}	e_{R61}	e_{R675}	e_{R364}	e_{R513}	e_{R417}		
1	+	+	+	+	+	+	+	+	e_2	
e_1	+	−	−	+	+	+	−	−	e_6	
e_2	−	−	−	−	+	+	+	+	1	
e_3	−	−	+	+	−	+	+	−	e_7	(ii.2)
e_4	−	+	−	+	−	+	−	+	e_5	
e_5	+	−	+	−	−	+	−	+	e_4	
e_6	−	+	+	−	+	+	−	−	e_1	
e_7	+	+	−	−	−	+	+	−	e_3	

	e_{R3}	e_{R41}	e_{R56}	e_{R72}	e_{R716}	e_{R475}	e_{R624}	e_{R521}		
1	+	+	+	+	+	+	+	+	e_3	
e_1	+	+	−	−	−	+	+	−	e_4	
e_2	+	−	−	+	+	+	−	−	e_7	
e_3	−	−	−	−	+	+	+	+	1	(ii.3)
e_4	−	−	+	+	−	+	+	−	e_1	
e_5	−	+	−	+	−	+	−	+	e_6	
e_6	+	−	+	−	−	+	−	+	e_5	
e_7	−	+	+	−	+	+	−	−	e_2	

	e_{R4}	e_{R52}	e_{R67}	e_{R13}	e_{R127}	e_{R516}	e_{R735}	e_{R632}		
1	+	+	+	+	+	+	+	+	e_4	
e_1	−	+	+	−	+	+	−	−	e_3	
e_2	+	+	−	−	−	+	+	−	e_5	
e_3	+	−	−	+	+	+	−	−	e_1	(ii.4)
e_4	−	−	−	−	+	+	+	+	1	
e_5	−	−	+	+	−	+	+	−	e_2	
e_6	−	+	−	+	−	+	−	+	e_7	
e_7	+	−	+	−	−	+	−	+	e_6	

	e_{R5}	e_{R63}	e_{R71}	e_{R24}	e_{R231}	e_{R627}	e_{R146}	e_{R743}	
1	+	+	+	+	+	+	+	+	e_5
e_1	+	−	+	−	−	+	−	+	e_5
e_2	−	+	+	−	+	+	−	−	e_7
e_3	+	+	−	−	−	+	+	−	e_4
e_4	+	−	−	+	+	+	−	−	e_6
e_5	−	−	−	−	+	+	+	+	1
e_6	−	−	+	+	−	+	+	−	e_3
e_7	−	+	−	+	−	+	−	+	e_1

(ii.5)

	e_{R6}	e_{R74}	e_{R12}	e_{R35}	e_{R342}	e_{R731}	e_{R257}	e_{R154}	
1	+	+	+	+	+	+	+	+	e_6
e_1	−	+	−	+	−	+	−	+	e_2
e_2	+	−	+	−	−	+	−	+	e_1
e_3	−	+	+	−	+	+	−	−	e_5
e_4	+	+	−	−	−	+	+	−	e_7
e_5	+	−	−	+	+	+	−	−	e_3
e_6	−	−	−	−	+	+	+	+	1
e_7	−	−	+	+	−	+	+	−	e_4

(ii.6)

	e_{R7}	e_{R15}	e_{R23}	e_{R46}	e_{R453}	e_{R142}	e_{R361}	e_{R265}	
1	+	+	+	+	+	+	+	+	e_7
e_1	−	−	+	+	−	+	+	−	e_5
e_2	−	+	−	+	−	+	−	+	e_3
e_3	+	−	+	−	−	+	−	+	e_2
e_4	−	+	+	−	+	+	−	−	e_6
e_5	+	+	−	−	−	+	+	−	e_1
e_6	+	−	−	+	+	+	−	−	e_4
e_7	−	−	−	−	+	+	+	+	1

(ii.7)

Appendix iii. $\mathbf{O}_L$ Actions: Product Rule $e_a e_{a+1} = e_{a+3}$.

Given the octonion multiplication,

$$e_a e_{a+1} = e_{a+3},$$

the resulting multiplication table is

$$\begin{bmatrix} 1 & e_1 & e_2 & e_3 & e_4 & e_5 & e_6 & e_7 \\ e_1 & -1 & e_4 & e_7 & -e_2 & e_6 & -e_5 & -e_3 \\ e_2 & -e_4 & -1 & e_5 & e_1 & -e_3 & e_7 & -e_6 \\ e_3 & -e_7 & -e_5 & -1 & e_6 & e_2 & -e_4 & e_1 \\ e_4 & e_2 & -e_1 & -e_6 & -1 & e_7 & e_3 & -e_5 \\ e_5 & -e_6 & e_3 & -e_2 & -e_7 & -1 & e_1 & e_4 \\ e_6 & e_5 & -e_7 & e_4 & -e_3 & -e_1 & -1 & e_2 \\ e_7 & e_3 & e_6 & -e_1 & e_5 & -e_4 & -e_2 & -1 \end{bmatrix}. \tag{iii.0}$$

The resulting adjoint actions are

$$\begin{array}{cccccccccc} & 1 & e_{L325} & e_{L436} & e_{L547} & e_{L651} & e_{L762} & e_{L173} & e_{L214} & \\ 1 & + & + & + & + & + & + & + & + & 1 \\ e_1 & + & - & - & - & + & - & + & + & e_1 \\ e_2 & + & + & - & - & - & + & - & + & e_2 \\ e_3 & + & + & + & - & - & - & + & - & e_3 \\ e_4 & + & - & + & + & - & - & - & + & e_4 \\ e_5 & + & + & - & + & + & - & - & - & e_5 \\ e_6 & + & - & + & - & + & + & - & - & e_6 \\ e_7 & + & - & - & + & - & + & + & - & e_7 \end{array} \tag{iii.1}$$

(iii.2)

	e_{L1}	e_{L37}	e_{L56}	e_{L24}	e_{L453}	e_{L764}	e_{L572}	e_{L632}	
1	+	+	+	+	+	+	+	+	e_1
e_1	−	−	−	−	+	+	+	+	1
e_2	+	−	−	+	+	+	−	−	e_4
e_3	+	+	−	−	−	+	+	−	e_7
e_4	−	+	+	−	+	+	−	−	e_2
e_5	+	−	+	−	−	+	−	+	e_6
e_6	−	+	−	+	−	+	−	+	e_5
e_7	−	−	+	+	−	+	+	−	e_3

(iii.3)

	e_{L2}	e_{L41}	e_{L67}	e_{L35}	e_{L564}	e_{L175}	e_{L613}	e_{L743}	
1	+	+	+	+	+	+	+	+	e_2
e_1	−	−	+	+	−	+	+	−	e_4
e_2	−	−	−	−	+	+	+	+	1
e_3	+	−	−	+	+	+	−	−	e_5
e_4	+	+	−	−	−	+	+	−	e_1
e_5	−	+	+	−	+	+	−	−	e_3
e_6	+	−	+	−	−	+	−	+	e_7
e_7	−	+	−	+	−	+	−	+	e_6

(iii.4)

	e_{L3}	e_{L52}	e_{L71}	e_{L46}	e_{L675}	e_{L216}	e_{L724}	e_{L154}	
1	+	+	+	+	+	+	+	+	e_3
e_1	−	+	−	+	−	+	−	+	e_7
e_2	−	−	+	+	−	+	+	−	e_5
e_3	−	−	−	−	+	+	+	+	1
e_4	+	−	−	+	+	+	−	−	e_6
e_5	+	+	−	−	−	+	+	−	e_2
e_6	−	+	+	−	+	+	−	−	e_4
e_7	+	−	+	−	−	+	−	+	e_1

	e_{L4}	e_{L63}	e_{L12}	e_{L57}	e_{L716}	e_{L327}	e_{L135}	e_{L265}	
1	+	+	+	+	+	+	+	+	e_4
e_1	+	−	+	−	−	+	−	+	e_2
e_2	−	+	−	+	−	+	−	+	e_1
e_3	−	−	+	+	−	+	+	−	e_6
e_4	−	−	−	−	+	+	+	+	1
e_5	+	−	−	+	+	+	−	−	e_7
e_6	+	+	−	−	−	+	+	−	e_3
e_7	−	+	+	−	+	+	−	−	e_5

(iii.5)

	e_{L5}	e_{L74}	e_{L23}	e_{L61}	e_{L127}	e_{L431}	e_{L246}	e_{L376}	
1	+	+	+	+	+	+	+	+	e_5
e_1	−	+	+	−	+	+	−	−	e_6
e_2	+	−	+	−	−	+	−	+	e_3
e_3	−	+	−	+	−	+	−	+	e_2
e_4	−	−	+	+	−	+	+	−	e_7
e_5	−	−	−	−	+	+	+	+	1
e_6	+	−	−	+	+	+	−	−	e_1
e_7	+	+	−	−	−	+	+	−	e_4

(iii.6)

	e_{L6}	e_{L15}	e_{L34}	e_{L72}	e_{L231}	e_{L542}	e_{L357}	e_{L417}	
1	+	+	+	+	+	+	+	+	e_6
e_1	+	+	−	−	−	+	+	−	e_5
e_2	−	+	+	−	+	+	−	−	e_7
e_3	+	−	+	−	−	+	−	+	e_4
e_4	−	+	−	+	−	+	−	+	e_3
e_5	−	−	+	+	−	+	+	−	e_1
e_6	−	−	−	−	+	+	+	+	1
e_7	+	−	−	+	+	+	−	−	e_2

(iii.7)

	e_{L7}	e_{L26}	e_{L45}	e_{L13}	e_{L342}	e_{L653}	e_{L461}	e_{L521}	
1	+	+	+	+	+	+	+	+	e_7
e_1	+	−	−	+	+	+	−	−	e_3
e_2	+	+	−	−	−	+	+	−	e_6
e_3	−	+	+	−	+	+	−	−	e_1
e_4	+	−	+	−	−	+	−	+	e_5
e_5	−	+	−	+	−	+	−	+	e_4
e_6	−	−	+	+	−	+	+	−	e_2
e_7	−	−	−	−	+	+	+	+	1

(iii.8)

Appendix iv. $\mathbf{O}_R$ Actions: Product Rule $e_a e_{a+1} = e_{a+3}$.

Finally

	1	e_{R325}	e_{R436}	e_{R547}	e_{R651}	e_{R762}	e_{R173}	e_{R214}	
1	$+$	$+$	$+$	$+$	$+$	$+$	$+$	$+$	1
e_1	$+$	$-$	$-$	$-$	$+$	$-$	$+$	$+$	e_1
e_2	$+$	$+$	$-$	$-$	$-$	$+$	$-$	$+$	e_2
e_3	$+$	$+$	$+$	$-$	$-$	$-$	$+$	$-$	e_3
e_4	$+$	$-$	$+$	$+$	$-$	$-$	$-$	$+$	e_4
e_5	$+$	$+$	$-$	$+$	$+$	$-$	$-$	$-$	e_5
e_6	$+$	$-$	$+$	$-$	$+$	$+$	$-$	$-$	e_6
e_7	$+$	$-$	$-$	$+$	$-$	$+$	$+$	$-$	e_7

(iv.0)

	e_{R1}	e_{R37}	e_{R56}	e_{R24}	e_{R453}	e_{R764}	e_{R572}	e_{R632}	
1	$+$	$+$	$+$	$+$	$+$	$+$	$+$	$+$	e_1
e_1	$-$	$-$	$-$	$-$	$+$	$+$	$+$	$+$	1
e_2	$-$	$+$	$+$	$-$	$-$	$-$	$+$	$+$	e_4
e_3	$-$	$-$	$+$	$+$	$+$	$-$	$-$	$+$	e_7
e_4	$+$	$-$	$-$	$+$	$-$	$-$	$+$	$+$	e_2
e_5	$-$	$+$	$-$	$+$	$+$	$-$	$+$	$-$	e_6
e_6	$+$	$-$	$+$	$-$	$+$	$-$	$+$	$-$	e_5
e_7	$+$	$+$	$-$	$-$	$+$	$-$	$-$	$+$	e_3

(iv.1)

	e_{R2}	e_{R41}	e_{R67}	e_{R35}	e_{R564}	e_{R175}	e_{R613}	e_{R743}	
1	+	+	+	+	+	+	+	+	e_2
e_1	+	+	−	−	+	−	−	+	e_4
e_2	−	−	−	−	+	+	+	+	1
e_3	−	+	+	−	−	−	+	+	e_5
e_4	−	−	+	+	+	−	−	+	e_1
e_5	+	−	−	+	−	−	+	+	e_3
e_6	−	+	−	+	+	−	+	−	e_7
e_7	+	−	+	−	+	−	+	−	e_6

(iv.2)

	e_{R3}	e_{R52}	e_{R71}	e_{R46}	e_{R675}	e_{R216}	e_{R724}	e_{R154}	
1	+	+	+	+	+	+	+	+	e_3
e_1	+	−	+	−	+	−	+	−	e_7
e_2	+	+	−	−	+	−	−	+	e_5
e_3	−	−	−	−	+	+	+	+	1
e_4	−	+	+	−	−	−	+	+	e_6
e_5	−	−	+	+	+	−	−	+	e_2
e_6	+	−	−	+	−	−	+	+	e_4
e_7	−	+	−	+	+	−	+	−	e_1

(iv.3)

	e_{R4}	e_{R63}	e_{R12}	e_{R57}	e_{R716}	e_{R327}	e_{R135}	e_{R265}	
1	+	+	+	+	+	+	+	+	e_4
e_1	−	+	−	+	+	−	+	−	e_2
e_2	+	−	+	−	+	−	+	−	e_1
e_3	+	+	−	−	+	−	−	+	e_6
e_4	−	−	−	−	+	+	+	+	1
e_5	−	+	+	−	−	−	+	+	e_7
e_6	−	−	+	+	+	−	−	+	e_3
e_7	+	−	−	+	−	−	+	+	e_5

(iv.4)

(iv.5)

	e_{R5}	e_{R74}	e_{R23}	e_{R61}	e_{R127}	e_{R431}	e_{R246}	e_{R376}	
1	+	+	+	+	+	+	+	+	e_5
e_1	+	−	−	+	−	−	+	+	e_6
e_2	−	+	−	+	+	−	+	−	e_3
e_3	+	−	+	−	+	−	+	−	e_2
e_4	+	+	−	−	+	−	−	+	e_7
e_5	−	−	−	−	+	+	+	+	1
e_6	−	+	+	−	−	−	+	+	e_1
e_7	−	−	+	+	+	−	−	+	e_4

(iv.6)

	e_{R6}	e_{R15}	e_{R34}	e_{R72}	e_{R231}	e_{R542}	e_{R357}	e_{R417}	
1	+	+	+	+	+	+	+	+	e_6
e_1	−	−	+	+	+	−	−	+	e_5
e_2	+	−	−	+	−	−	+	+	e_7
e_3	−	+	−	+	+	−	+	−	e_4
e_4	+	−	+	−	+	−	+	−	e_3
e_5	+	+	−	−	+	−	−	+	e_1
e_6	−	−	−	−	+	+	+	+	1
e_7	−	+	+	−	−	−	+	+	e_2

(iv.7)

	e_{R7}	e_{R26}	e_{R45}	e_{R13}	e_{R342}	e_{R653}	e_{R461}	e_{R521}	
1	+	+	+	+	+	+	+	+	e_7
e_1	−	+	+	−	−	−	+	+	e_3
e_2	−	−	+	+	+	−	−	+	e_6
e_3	+	−	−	+	−	−	+	+	e_1
e_4	−	+	−	+	+	−	+	−	e_5
e_5	+	−	+	−	+	−	+	−	e_4
e_6	+	+	−	−	+	−	−	+	e_2
e_7	−	−	−	−	+	+	+	+	1

Bibliography

[1] I.R. Porteous, *Topological Geometry*, (Cambridge, 2nd Ed., 1981).

R. Ablamowicz, P. Lounesto, J. Maks: Report on the Second "Workshop on Clifford Algebras and Their Applications in Mathematical Physics," Found. Phys. **21** (1989) 201.

[2] G.M. Dixon, Algebraic Unification, Phys. Rev. D **28** (1983) 833.

[3] G.M. Dixon, Generalized Clifford Algebras: Orthogonal and Symplectic Cases, Lett. Math. Phys. **5** (1981) 411.

[4] G.M. Dixon, Fermionic Clifford Algebras and Supersymmetry, *Clifford Algebras and their Application in Mathematical Physics*, (D. Reidel Publishing Co., 1986) 393.

[5] V.G. Kac, A Sketch of Lie Superalgebra Theory, Comm. Math. Phys. **53** (1977) 31.

[6] G.M. Dixon, Derivation of the Standard Model, Il Nuovo Cimento **105B** (1990) 349.

[7] C.A. Manogue, A. Sudbery, General Solutions of Covariant Superstring Equations of Motion, Phys. Rev. D **40** (1989) 4073.

C.A. Manogue, J. Schray, Finite Lorentz Transformations, Automorphisms, and Division Algebras, J. Math. Phys. **34** (1993) 3746.

[8] I. Bengtsson, M. Cederwall, Particles, Twistors and Division Algebras, Nuc. Phys. **B302** (1988) 81.

M. Cederwall, C.R. Preitschopf, S^7 and $\hat{S}^7$, hep-th-9309030.

[9] M. Günaydin, G. Sierra, P.K. Townsend, Exceptional Supergravity Theories and the Magic Square, Phys. Lett. **133B** (1983) 72.

[10] J.H. Conway, N.J.A. Sloane, *Sphere Packings, Lattices and Groups*, (Springer-Verlag, 2nd Ed., 1991).

[11] F.D. Smith, Calculation of 130 GeV Mass for T-quark, hep-ph-9301210.

F.D. Smith, $SU(3) \times SU(2) \times U(1)$, Higgs, and Gravity from $Spin(0,8)$ Clifford Algebra $Cl(0,8)$, hep-th-9402003.

[12] E.S. Abers, B.W. Lee, Phys. Rep. **9C** (1973) 1.

[13] P. Candelas, G.T. Horowitz, A. Strominger, E. Witten, Superstring Phenomenology, *Symposium on Anomolies, Geometry, Topology*, (World Scientific Publishing Co. Pte. Ltd., 1985) 377.

[14] A. Sudbery, Division Algebras, (Pseudo)orthogonal Groups and Spinors, J. Phys. A **17** (1984) 939.

[15] M. Günaydin, G. Sierra, P.K. Townsend, The Geometry of N=2 Maxwell-Einstein Supergravity and Jordan Algebras, Nuc. Phys. **B242** (1984) 244.

Index

Other *Mathematics and Its Applications* titles of interest:

F. Langouche, D. Roekaerts and E. Tirapegui: *Functional Integration and Semiclassical Expansions.* 1982, 328 pp. ISBN 90-277-1472-X

N.E. Hurt: *Geometric Quantization in Action. Applications of Harmonic Analysis in Quantum Statistical Mechanics and Quantum Field Theory.* 1982, 352 pp. ISBN 90-277-1426-6

C.P. Bruter, A. Aragnal and A. Lichnerowicz (eds.): *Bifurcation Theory, Mechanics and Physics. Mathematical Developments and Applications.* 1983, 400 pp. *out of print,* ISBN 90-277-1631-5

N.H. Ibragimov: *Transformation Groups Applied to Mathematical Physics.* 1984, 414 pp. ISBN 90-277-1847-4

V. Komkov: *Variational Principles of Continuum Mechanics with Engineering Applications. Volume 1: Critical Points Theory.* 1986, 398 pp. ISBN 90-277-2157-2

R.L. Dobrushin (ed.): *Mathematical Problems of Statistical Mechanics and Dynamics. A collection of Surveys.* 1986, 276 pp. ISBN 90-277-2183-1

P. Kree and C. Soize: *Mathematics of Random Phenomena. Random Vibrations of Mechanical Structures.* 1986, 456 pp. ISBN 90-277-2355-9

H. Triebel: *Analysis and Mathematical Physics.* 1987, 484 pp. ISBN 90-277-2077-0

P. Libermann and Ch.M. Marle: *Sympletic Geometry and Analytical Mechanics.* 1987, 544 pp. ISBN 90-277-2438-5 (hb), ISBN 90-277-2439-3 (pb).

E. Tirapegui and D. Villarroel (eds.): *Instabilities and Nonequilibrium Structures.* 1987, 352 pp. ISBN 90-277-2420-2

W.I. Fushchich and A.G. Nikitin: *Symmetries of Maxwell's Equations.* 1987, 228 pp. ISBN 90-277-2320-6

V. Komkov: *Variational Principles of Çontinuum Mechanics with Engineering Applications. Volume 2: Introduction to Optimal Design Theory.* 1988, 288 pp. ISBN 90-277-2639-6

M.J. Vishik and A.V. Fursikov: *Mathematical Problems in Statistical Hydromechanics.* 1988, 588 pp. ISBN 90-277-2336-2

V.I. Fabrikant: *Applications of Potential Theory in Mechanics. A Selection of New Results.* 1989, 484 pp. ISBN 0-7923-0173-0

E. Tirapegui and D. Villarroel (eds.): *Instabilities and Nonequilibrium Structures (II). Dynamical Systems and Instabilities.* 1989, 328 pp. ISBN 0-7923-0144-7

V.G. Bagrov and D.M. Gitman: *Exact Solutions of Relativistic Wave Equations.* 1990, 344 pp. ISBN 0-7923-0215-X

Other *Mathematics and Its Applications* titles of interest:

O.I. Zavialov: *Renormalized Quantum Field Theory.* 1990, 560 pp.
ISBN 90-277-2758-9

S.S. Horuzhy: *Introduction to Algebraic Quantum Field Theory.* 1990, 360 pp.
ISBN 90-277-2722-8

V.G. Makhankov: *Soliton Phenomenology.* 1990, 461 pp. ISBN 90-277-2830-5

S. Albeverio, PH. Blanchard and L. Streit: *Stochastic Processes and their Applications in Mathematics and Physics.* 1990, 416 pp. ISBN 0-9023-0894-8

V.A. Malyshev and R.A. Minlos: *Gibbs Random Fields. Cluster Expansions.* 1991, 248 pp. ISBN 0-7923-0232-X

E. Tirapegui and W. Zeller (eds.): *Proceedings of the 3rd Workshop on Instabilities and Nonequilibrium Structures.* 1991, 370 pp. ISBN 0-7923-1153-1

C.A. Marinov and P. Neittaanmaki: *Mathematical Models in Electrical Circuits. Theory and Applications.* 1991, 160 pp. ISBN 0-7923-1155-8

C. Bartocci, U. Bruzzo and D. Hernandez-Ruiperez: *The Geometry of Supermanifolds.* 1991, 242 pp. ISBN 0-7923-1440-9

E. Goles and S. Martinez: *Statistical Physics, Automata Networks and Dynamical Systems.* 1992, 208 pp. ISBN 0-7923-1595-2

A. van der Burgh and J. Simonis (eds.): *Topics in Engineering Mathematics.* 1992, 266 pp. ISBN 0-7923-2005-3

L. Aizenberg: *Carleman's Formulas in Complex Analysis.* 1993, 294 pp.
ISBN 0-7923-2121-9

E. Tirapegui and W. Zeller (eds.): *Instabilities and Nonequilibrium Structures IV.* 1993, 371 pp. ISBN 0-7923-2503-6

G.M. Dixon: *Division Algebras: Octonions, Quaternions, Complex Numbers and the Algebraic Design of Physics.* 1994, 246 pp. ISBN 0-7923-2890-6

MIX
Papier aus verantwortungsvollen Quellen
Paper from responsible sources
FSC® C105338

If you have any concerns about our products,
you can contact us on
ProductSafety@springernature.com

In case Publisher is established outside the EU,
the EU authorized representative is:
Springer Nature Customer Service Center GmbH
Europaplatz 3, 69115 Heidelberg, Germany

Printed by Libri Plureos GmbH
in Hamburg, Germany